BBC DOCTOR WHO

THE TIME LORD LETTERS

JUSTIN RICHARDS

HARPER DESIGN
An Imprint of HarperCollins Publishers

DOCTOR WHO: THE TIME LORD LETTERS
Text © Justin Richards 2015

Doctor Who is a BBC Wales production for BBC One.
Executive producers: Steven Moffat and Brian Minchin

HarperCollins books may be purchased for educational, business, or sales promotional use. For information please e-mail the Special Markets Department at SPsales@harpercollins.com.

First published in 2015 by:
Harper Design
An *Imprint of HarperCollinsPublishers*
195 Broadway
New York, NY 10007
Tel: (212) 207-7000
Fax: 855-746-6023
harperdesign@harpercollins.com
www.hc.com

This edition distributed throughout the world by:
HarperCollins*Publishers*
195 Broadway
New York, NY 10007

Library of Congress Control Number: 2015939503

ISBN 978-0-06-239728-7

First Printing, 2015

Editorial director: Albert DePetrillo
Editor: Charlotte Macdonald
Series consultant: Justin Richards
Design: Amazing 15
Production: Phil Spencer

With thanks to Tom Spilsbury, Peter Ware, and all at *Doctor Who Magazine*

All images © BBC

All design furniture © Shutterstock

Printed and bound in China by Toppan Leefung

Contents

INTRODUCTION

RESEARCHING THE LIVES AND TIMES OF THE DOCTOR IS DIFFICULT. WHILE HE HAS BEEN INVOLVED IN PIVOTAL EVENTS IN OUR HISTORY AND IN MANY OTHER PEOPLE'S HISTORIES - AND HAS EVEN BEEN RESPONSIBLE FOR SOME OF THEM - INFORMATION ABOUT THE DOCTOR IS SURPRISINGLY SCARCE. HE IS NOT A MAN WHO ENJOYS THE LIMELIGHT OR WHO PUBLICISES HIS ACTIONS. INDEED, HE ONCE ATTEMPTED TO ERASE ALL MENTION OF HIMSELF FROM THE ENTIRE INTERNET, AND HAS ALSO DELETED HIMSELF FROM MOST DATABASES IN THE UNIVERSE.

But no one could possibly travel through history - past, present and future - as much as the Doctor does without leaving an impression. Much of what we know about this enigmatic traveller comes from his actions, from the planets he has saved and the monsters he has defeated. But we also have another source of information - his writings.

The most famous written work by the Doctor was a short message blasted onto a ruined wall at the fall of the Gallifreyan city of Arcadia towards the end of the Last Great Time War. The words 'No More' took on a grim relevance shortly afterwards when the Doctor himself brought the Time War to an abrupt end. But new research has afforded us access to many more documents either written by or addressed to the Doctor.

As well as providing brief background information on relevant subjects, such as the Doctor's home planet Gallifrey, this book offers a unique collection of over one hundred letters, notes and jottings both by and to the Doctor. They span centuries of history and future history as well as light years of space. They vary from the serious, the emotional and the grim to the flippant, the funny and the incredible.

Brought together from archives and collections across the universe and throughout time, this unique collection sheds new light on one of the most mysterious figures in all of time and space.

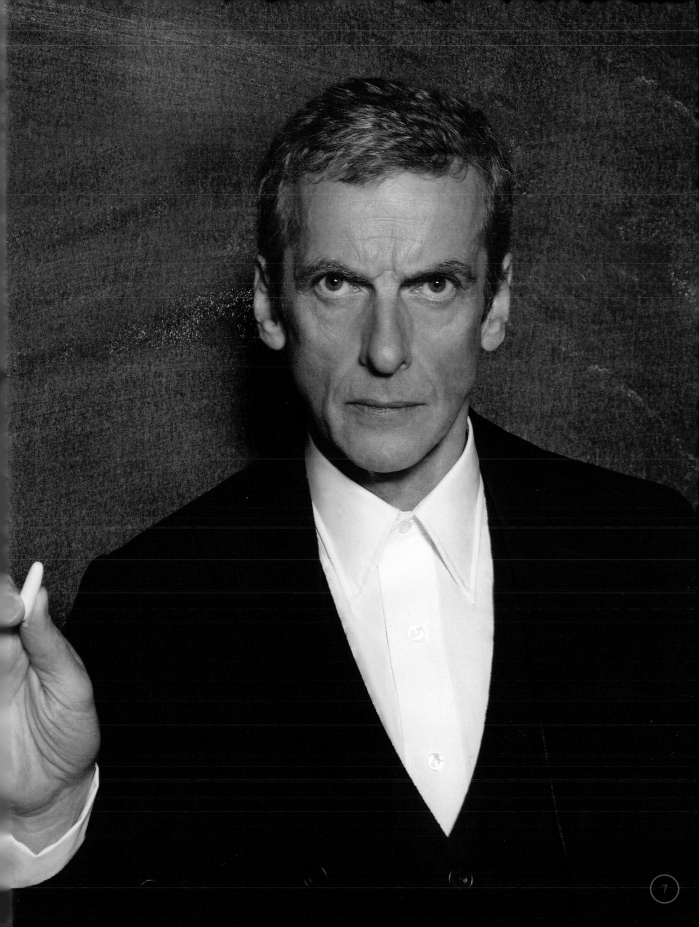

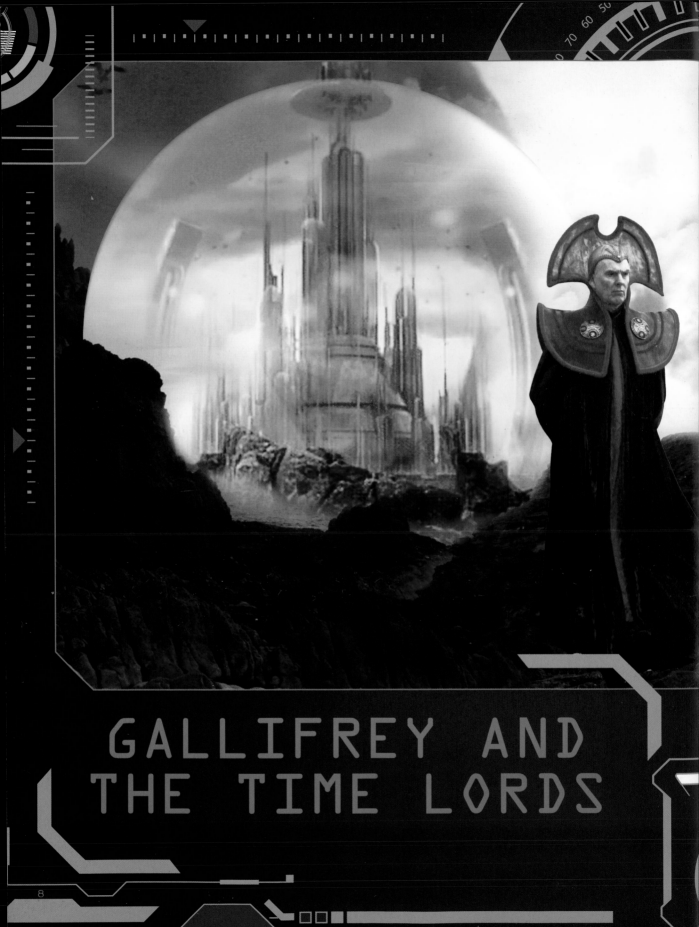

GALLIFREY AND
THE TIME LORDS

GALLIFREY - THE SHINING WORLD OF THE SEVEN SYSTEMS - WAS HOME TO THE ANCIENT AND POWERFUL TIME LORDS. IN THE OLD TIME, LED BY THE ALMOST MYTHICAL FIGURE OF RASSILON, THEY DISCOVERED THE SECRETS OF SPACE-TIME TRAVEL. IT WAS THE STELLAR ENGINEER OMEGA WHO PROVIDED THE IMMENSE POWER NEEDED TO LEND A PRACTICAL APPLICATION TO RASSILON'S THEORIES.

As a result of their disastrous early involvement in the affairs of other races, like the Minyans, the Time Lords evolved a policy of non-intervention. They would observe the universe, but not get involved in it. Some Time Lords felt this was a mistake, and wanted to play a more active role in the universe - exploring and experiencing other times and places. One of these was the Doctor.

In his youth, when he left Gallifrey, the Doctor was a much older man. The world he left behind held many memories for him, with twin suns shining in a burnt orange sky. It was on Gallifrey that he attended Prydon Academy. Tutored by Cardinal (and future President) Borusa the Doctor graduated as a full member of the Prydonian Chapter. Amongst his contemporaries were the Master and the Rani, both of whom also left Gallifrey and both of whom the Doctor would encounter again...

The Doctor's early travels were unplanned and largely random owing to the fact that he had stolen a TARDIS with a faulty guidance system. But this was also to his advantage, as it meant the Time Lords were unable to track him down and bring him back to Gallifrey. It was not until he was forced by circumstance to call on his own people for help that they found him. When they did, the Doctor was put on trial for his crime - for interfering in the affairs of others.

His sentence was, however, relatively lenient - an enforced change of appearance and exile to Earth in the late twentieth century. When his exile was lifted, following the Doctor's help against Omega, he resumed his travels. But now he was often at the beck and call of the Time Lords, and even returned to Gallifrey on several occasions. It was, after all, his home.

Until the Last Great Time War...

Some say the Doctor was involved in starting the War - sent by the Time Lords to avert the creation of the Daleks. He failed, but millennia later it was the Doctor who brought the War to its terrible end. For a while it seemed the Doctor was the only survivor - the last of the Time Lords.

Ravaged and left desolate, Gallifrey was believed destroyed in the Last Great Time War between the Time Lords and the Daleks. But rumours persist that it may somehow have survived. If it has, it can only be a matter of time before the Doctor finds it.

One of the earliest items of correspondence from the Doctor's life, this letter dates back to his school days. Written in response to his Interim Academy Report, the Doctor - using his Academy Student Identification Code 'Theta Sigma' - sent this reply to his personal tutor, Borusa.

Borusa was at the time the head of Prydon Academy, the college from which all Prydonians, like the Doctor, graduated. He later rose to the rank of Cardinal, before being elevated to the High Council. He eventually took over as President because the duly elected President was absent. The duly elected President was in fact the Doctor, who had actually been elected by default when the only other candidate was killed. He later resigned in favour of Borusa...

This letter is reproduced with kind permission of the Academy Archives, Gallifrey.

Dear Borusa

Thank you for the Interim Report on my progress which appeared in
my student correspondence data today. I have read it with interest,
and while I appreciate there are indeed some areas of my academic
and social progression that could perhaps be enhanced, I feel I must
take issue with some of the comments and observations you and your
fellow tutors have made.

The suggestion that I have a 'propensity for vulgar facetiousness' I
found particularly hurtful, especially as it is not backed up with
any real evidence. I do admit that the incident with the perigosto
stick and the temporal feedback loop was perhaps not in the best
of taste, but I have already appeared before the Academy Board
to explain my reasoning there and to demonstrate the interesting
findings of this experiment. And it's not as if High Tutor Albrecht
didn't have a few regenerations to spare.

I note also that you have marked down my work in Temporal
Engineering. I fail to see why this was necessary. I did, as I'm
sure you will recall, resubmit my Random Field Frame paper, revised
as you suggested to avoid any reference to the Shaboogans or the
Minyan Incident. Regrading it to Epsilon Minor merely because of
what you call 'tonal problems' seems rather unfair. I should point
out that while sarcasm is perceived by some to be the lowest form of
wit, it is also one of the most effective.

I won't go into detail about my feelings towards your general
comments, or the suggestion that 'he might as well have been
brought up in a barn rather than a fully techno-temporal society.'
I think the lack of appreciation for social differentiation and
individualism this remark demonstrates speaks for itself.

Instead, I shall end by saying that if it were up to the students to
report back on their tutors rather than vice versa, then I struggle
to think of a single member of the Faculty (with the possible
exception of Lord Azmael) who would not come away with a rating of
'Could Try Harder'. Surely the progress and success of the pupils is
a reflection on their teachers.

I look forward to seeing you in the next Gravitational Induction
seminar.

Theta Sigma

This handwritten note was discovered in the Time Capsule Storage Bays beneath the Citadel of Gallifrey. The technician who found it immediately informed the Castellan, but it was too late to stop the missing Type 40 TARDIS capsule from leaving Gallifrey. In fact, it was not entirely clear when the note was left or how long the TARDIS had been gone. Since the missing capsule was in the Repair Section, some considerable time could have elapsed before the loss was noticed.

A subsequent enquiry found that the Transduction Barriers had been disabled before the capsule was removed from storage. The system's Fault Locator had been rewired so that no indication of a malfunction showed on the monitoring systems.

150 mm
50 mm
18 mm

POLICE   PUBLIC CALL   BOX

To Whom It May Concern

As you have no doubt noticed, I have removed a TARDIS Time Capsule from this location. I do not intend to return it. As it is a rather old model and was in for repair, I really cannot imagine that anyone will miss it.

But assuming someone does, and an explanation is required, allow me to elucidate. In short, I have had enough of monitoring and analysis. I am bored with watching rather than doing. With a whole universe of space and time to explore, all we Time Lords ever do is observe. Well, I don't know about you, but for Susan and myself, this is no longer enough.

Should you wish the TARDIS returned, you will find me somewhere in space and time. But, quite frankly, we would rather not be bothered.

The Doctor

1535 mm

In 1066, shortly before the Battle of Hastings, the First Doctor encountered another Time Lord, who was masquerading as a monk in an abandoned monastery in Northumbria. The 'Monk' was planning to adjust history so that King Harold would defeat William the Conqueror. The Doctor managed to prevent the Monk from changing the course of history, and sabotaged his TARDIS - disguised as a Saxon sarcophagus - so that its interior dimensions shrank to the size of its external shell.

Although unable to get back inside his shrunken TARDIS (at least until he later repaired it), the Monk was able to retrieve this note left for him by the Doctor.

My dear fellow

I'm sure will you excuse me but I didn't want to say goodbye, as you are obviously going to be very busy for some time. Just in case you still have ideas about your master plan, I've taken precautions to stop your time meddling. Possibly one day in the future, when you've learnt your lesson, I shall return and release you.

I'm sure that even you will soon notice that I have removed your TARDIS's Dimensional Controller. I hope that will be a lesson to you about your tiresome meddling.

Quite apart from the irresponsible childishness of wanting Harold to win at Hastings simply because in your opinion he would have been a better king than William, you seem to have no idea of the possible implications and repercussions of your actions. It's one thing to wear a wristwatch in the eleventh century, and avail yourself of the relative comforts of a toaster, a stove, a box of snuff and a gramophone. But your 'collection' of paraphernalia from all the times and places you have visited shows just how juvenile your meanderings have become.

I should also point out that if you are going to flout the Laws of Time so flagrantly, it is perhaps not the most sensible idea to keep notes of all your various transgressions. Discussing the principles of powered flight with Leonardo is nothing to be proud of. Nor is depositing £200 in a bank in 1968 then 'nipping forward 200 years' as you put it to collect a fortune in compound interest.

As for your use of an anti-gravitational lift to help the Ancients build Stonehenge, well I hardly know where to begin.

So, as I say, I trust that this will be a lesson to you and that this irresponsible meddling will cease forthwith.

Yours etc.

The Doctor

Having kept well away from his own people and done everything he could to avoid them after stealing the TARDIS, the Doctor was finally forced to contact the Time Lords. He had defeated an alien War Lord's plans to create an army out of human soldiers taken from various time zones on Earth. But now he was faced with a problem. As the aliens' own time machines had stopped working, he had no way of returning the humans to their own times and locations.

Hoping to be able to escape before the Time Lords arrived, the Doctor sent them a telepathic message – a transcript of which is preserved in the Panopticon Archives.

//// You know I wouldn't contact you unless it was absolutely necessary and there was no alternative. But I need your help. There, I've said it – I need your help.

//// To be honest, I'm not even sure what planet I'm on, let alone the real time period. But stranded here are thousands – possibly millions – of innocent human beings who have been taken out of their own time by callous aliens and made to fight against each other in fabricated environments. You can see the extent of the operation from the thought-images I have included with this message.

//// And before you turn away and decide this is nothing to do with you, let me tell you that it is. You might want to wash your hands of it, to claim that you never get involved, but the time travel technology the aliens used was provided by one of us - by a Time Lord. We helped create this problem, and I believe we are obliged to sort it out.

//// I have done what I can. I have ended the War Games and stopped the killing. But now - and I say it again - I need your help.

//// I just hope that by the time you arrive and sort out this terrible situation, I shall be long gone. But if not, then my continued freedom is a small thing to give up in exchange for the unfulfilled lives of the humans trapped here.

Apprehended by the Time Lords, the Doctor was put on trial for breaking one of their cardinal rules - the law that forbids Time Lord interference in the affairs of other planets and peoples. As well as presenting his defence to a triumvirate of Time Lord judges, the Doctor submitted a written deposition in which he made the case for the Time Lords becoming more involved in the affairs of other worlds.

This is an extract from that deposition, reference Malfeasance Tribunal Hearing dated 3-0-9-9-0-6, reproduced with kind permission of the Panopticon Archives.

behaviour that you deem to be illegal, I contend is nothing to be ashamed of. Indeed, I am proud of my actions. Should we Time Lords, with all our power and insight, be content merely to observe the evil in the universe? I believe the answer to this is a resounding 'No'. Rather, we should be fighting against it.

When I present my defence I shall further justify my actions by showing the Tribunal some of the evils I have been fighting against. If permitted access to a thought channel, I shall happily reveal the cruelty dealt out by the creatures like the Quarks – deadly robot servants of the cruel Dominators who tried to enslave the peace-loving people of Dulkis. I shall tell of the Ice Warriors, cruel Martian invaders who tried to conquer Earth. I will defend my actions against the Cybermen – a race half creature, half machine that are determined to survive at any cost.

And of course, you already know about the Daleks – a pitiless race of conquerors exterminating all who come up against them. Yet you do nothing to protect and defend the lesser races that are their victims. I may be guilty of interference, but you – the Time Lords – are just as guilty of failing to use your great powers to help those in need.

I believe that we can no longer deny that there is evil in the universe that must be fought. And with my experience and first-hand knowledge of that evil, I have a part to play in that battle.

But even if you reject my arguments, even if you continue to deny what must be obvious, I plead for leniency not for myself but for my companions Jamie and Zoe. They know nothing of the Time Lords, or of your laws. Whatever crimes they have committed, they have done so unwittingly and under my auspices. Whatever fate awaits me, it should not be visited upon these two innocent human beings

After defeating the Kastrian criminal Eldrad, the Doctor was summoned back to Gallifrey for a very special ceremony - the Resignation of the President. As well as other resignation honours, it was in the gift of the departing President to name his or her successor to the position. Usually, but not always, the serving Chancellor was named as the next President.

But as well as the summons back to the Panopticon, the Doctor also had a vision - a premonition: he saw the President assassinated as he prepared to give his resignation speech and name his successor.

Arriving back on Gallifrey, the Doctor found that his TARDIS was immediately surrounded by officers of the Chancellery Guard, led by Commander Hilred. Determined to try to save the President, the Doctor reasoned he could only do so if he avoided arrest. But he also wanted to warn the authorities of the possible threat.

Reproduced here is the handwritten note that the Doctor left in his TARDIS warning of the plot to kill the President.

TRANSLATION:

///TO THE CASTELLAN OF THE CHANCELLERY GUARD

/// I HAVE GOOD REASON TO THINK THAT THE LIFE OF HIS SUPREMACY THE PRESIDENT IS IN GRAVE DANGER. DO NOT IGNORE THIS WARNING.

/// THE DOCTOR

This message was received by the Coordinator of the Panopticon Archives. It was sent shortly after the Doctor, newly regenerated into his sixth incarnation, defeated the plans of the Giant Gastropod Mestor.

//// To Whom It May Concern

//// It is with the deepest of regret and acute personal sadness that I have to report the death of a fellow Time Lord.

//// Lord Azmael was one of the more accessible and amenable – not to say learned – of my tutors at Prydon Academy. It is largely because of his influence and teachings, along with his colleague Cardinal Borusa, that I graduated and am proud to regard myself as a Prydonian.

//// For all his wisdom and learning, Azmael never forgot what it was like to be a student. Given the chance, I am sure he would have continued in academia all his lives. When I visited him many years (and several regenerations) after I left the Academy, he and I spent a very pleasant evening reminiscing. An evening which only ended when I had to throw him into the Fountain of Vermillitan in the main square to sober him up again.

//// As you will be aware, Azmael was one of the few Time Lords well-regarded enough to be permitted to retire from Gallifrey. After some time continuing his studies and increasing still further his phenomenal knowledge, he became the Master of Jaconda. And it was on this benighted planet that he met his tragic end.

//// For a while he fell under the malign influence of Mestor, leader of the giant Gastropods that subjugated Jaconda. But eventually, with my help of course, he was able to break through this mental conditioning, and oppose Mestor. Together we brought a halt to the creature's plan to trigger a solar eruption that would scatter Gastropod eggs across the universe, destroying Jaconda in the process.

//// The planet Jaconda - and the peoples of the entire universe - owe Lord Azmael a debt of thanks that can never be repaid.

//// I ask that his name be entered into the Gallifreyan Book of Heroes, and that his surviving Data Extract be uploaded to the APC Net of the Matrix so that he may in some fashion live on and be remembered.

//// Your humble servant

//// The Doctor

A rather different tone is evident in this message, also from the Panopticon Archives. The message was sent by the Third Doctor using his TARDIS telepathic circuits. It originated in the 26th century time zone from a largely uninhabited sector on the edge of the galaxy.

//// Yes, well, I'm no more happy about this than you are. But I suppose there's no other option. All right then...

//// I don't have long as I'm close to collapse, so let's make this quick. I need you to do something for me for a change. I need you to direct the TARDIS. I know I usually complain when you do that, and demand that you stop treating me like a galactic yo-yo, but this time it's different.

//// Daleks.

//// Not without some difficulty, I have just managed to foil a plot by the Master to start a war between the empires of Earth and Draconia. You probably know that already if you're monitoring my timelines. Which seems likely.

//// But it turns out that the Master was working for the Daleks. Probably intending to turn on them at some point when he'd got them to do his dirty work, which is risky to say the least. That sort of behaviour is what will get him killed one of these days. I'm sorry to admit that when it does I shall rather miss him.

//// But I'm straying off the point – which is this: the Daleks are heading back from the Ogrons' homeworld to wherever they've got their army waiting. I imagine this warmongering nonsense is merely a part of their plan. So I need you to send the TARDIS after the Dalek ship.

//// That's all. Nothing strenuous. I'll do all the difficult stuff to stop the Daleks when I get there. Wherever 'there' turns out to be.

//// I'm sorry, I'll have to go now. Feeling a bit woozy…

RECEIVED AND ACTIONED – BY ORDER OF CASTELLAN TERRYNATE, REFERENCE 7-4-73/SSS/12-5-73

Even after the Doctor's exile was ended in recognition of his help in dealing with the renegade Omega, the Time Lords continued to rely on the Doctor. Sometimes, they explicitly asked for his assistance - as when they sent him back to the end of the Thal-Kaled war on the planet Skaro in an effort to avert the creation of the Daleks. But on other occasions they simply redirected the TARDIS to deliver the Doctor into a situation they wanted him to deal with.

If the Time Lords hoped the Doctor would not realise what was happening, they were sadly mistaken. Following their redirection of the TARDIS to the planet Karn, where the executed Time Lord President and war criminal Morbius was being reconstituted in a new body by the noted surgeon Mehendri Solon, the Doctor sent the following message to the High Council. Unusually, it was handwritten and delivered by trans-portal rather than being sent telepathically.

You could just ask, you know.

For a race that apparently prides itself on not interfering in the affairs of others, you seem to take a great deal of pleasure in continually interfering with my life. Whatever I've done in the past, I can't help feeling I've more than made up for it. I've served my sentence — and paralysingly dull, boring and tedious it was too.

So rather than just diverting my TARDIS to some godforsaken rock in the hope that I'll do your dirty work and save you dirtying your lily-white hands, why not ask nicely. I mean, if you'd actually told me that the brain of Morbius had survived his execution and a mad scientist was currently stitching together a new body for it out of the remains of spaceship crash victims, then I might well have been inclined to do something about it.

But oh no. You let me blunder in, upset the Sisterhood of Karn — which for the record is not a good idea — and nearly get poor Sarah killed into the bargain.

So can you please just stop?

Because if you carry on like this I'm likely to get angry. And if I'm angry I'm less likely to help.

Yours in irritation

The Doctor

Following his capture after putting an end to the cruel War Games, the Doctor was put on trial by the Time Lords. But that was not the last time he was brought before a Time Lord court.

In his sixth incarnation, the Doctor was indicted by the Valeyard - who was actually an aspect of the Doctor himself - and held to account before Inquisitor Darkel in an inquiry that later became a trial.

During the course of events, it became clear that the evidence had been manipulated by the Valeyard, and the Doctor was ultimately acquitted and discharged. Reproduced here is the final document from the trial proceedings - a statement from the Doctor, written after the trial had ended, and leaving the Time Lords in no doubt as to his feelings about the incident.

It's difficult knowing where to start. I don't know what riles me more – being abducted from my time line, being put on trial for spurious, trumped-up charges, watching my friend die in what then turned out to be an illusory temporal replay, or discovering that I have in my absence from Gallifrey been deposed as Lord President. Without anyone actually bothering to tell me about it.

No doubt some of this was intended to teach me some sort of lesson. It hasn't. In fact, it's more than apparent that it should be you, the Time Lords, who learn lessons.

Quite apart from allowing yourselves to be duped by the Valeyard – in a convoluted but rather obvious plan that even the Master (a) wasn't fooled by and (b) had to explain to you – you almost succeeded in allowing this Machiavellian misfit to destroy all of Time Lord society. Congratulations.

Perhaps now you will begin to appreciate the help and advice I have modestly offered over the millennia. Perhaps now you will begin to under-stand why experience of the real universe is useful and productive and not something rather strange and eccentric. Perhaps now you will begin to put your own house in order.

I said it at the farce that was my so-called trial, and I shall say it again now: In all my travels throughout the universe I have battled against evil, against power-mad conspirators. I should have stayed here. The oldest civil-isation – decadent, degenerate and rotten to the core. Power-mad conspira-tors, Daleks, Sontarans, Cybermen, they're still in the nursery compared to us Time Lords. Ten million years of absolute power, that's what it takes to be really corrupt.

You seem to have decided you don't want me as President, and that's fine. But at least appoint someone who can sort it out. Or if you can't find such a person on Gallifrey, then you know where I am.

The Doctor

PLANET EARTH

IF THERE IS ONE PLANET WHICH THE DOCTOR CAN CALL HIS HOME AS MUCH AS, IF NOT MORE THAN GALLIFREY, THAT PLANET IS EARTH. IN FACT, DURING HIS EXILE, EARTH REALLY WAS THE DOCTOR'S HOME. BUT THE TIME LORDS CHOSE TO SEND THE DOCTOR THERE IN THE FIRST PLACE BECAUSE THEY HAD NOTED HIS 'PARTICULAR INTEREST' IN THE PLANET, DECIDING THAT 'THE FREQUENCY OF YOUR VISITS MUST HAVE GIVEN YOU SPECIAL KNOWLEDGE OF THAT WORLD AND ITS PROBLEMS.'

For all he vehemently complained about being stuck on one primitive planet in one short period of time, the Doctor was slow to sever his ties once his exile was lifted. Although his visits became gradually more infrequent, he continued to work as UNIT's Scientific Adviser for a while. It is a role he has never really escaped from. But even though he spent less time with UNIT, the Doctor spent just as much time on Earth as he ever did. Much later, Romana would apparently surprise the Doctor with the news that 'everyone knows' Earth is his favourite planet.

In the same way, humans are - in the Doctor's own words - 'quite my favourite species'. Indeed, the vast majority of the Doctor's travelling companions and friends have come from Earth, or are of human descent. Initially this was accidental - two of his granddaughter's school teachers stumbling into the TARDIS - but later it was quite deliberate. Once the Doctor was happy to return his companions to their homes, or find them new ones, and then dematerialise from their lives for good. But now he seems to relish returning to visit Amy and Rory, or Clara.

Planet Earth itself seems to be more vulnerable to alien threats than other planets, as the Second Doctor noted in his trial. We know that the planet's position is strategically significant - and that may be one reason why it has been the target of so many attempted invasions.

But whatever the reason for the Doctor's continuing fascination with our planet, the sheer number of times he has visited throughout its history, and the number of people he has met and influenced, makes Earth as significant to the Doctor as the Doctor has been to Earth.

This handwritten letter was delivered to King Henry VIII in October 1543. It had been left in a cell in the Tower of London by the prisoner locked inside. According to the Tower records, although the door was locked and there was no other means of escape, the two prisoners - an elderly man and his young granddaughter - had gone from the cell. Also missing was a large blue cabinet which Henry had ordered stored in the same cell.

(Note that a 'Parson's Nose' is a part of dressed and cooked poultry such as a chicken or turkey.)

Your Majesty

I thought it only right and correct to write to you and convey my thanks for this incarceration. I have to say that the amenities are somewhat antiquated, even for the sixteenth century, but Susan and I do not intend staying long.

I suppose I should apologise for quarrelling with you at dinner. It was in all other respects quite a fine repast. I do not blame you at all for losing your temper and throwing that parson's nose at me. Nor do I apologise for throwing it back at you. Quite apart from the crass nature of your comments and your rather childish behaviour, I had another motive. Though I am forced to wonder if you are really mature enough in your outlook to have been elevated to such an exalted post as monarch.

But be that as it may, you have done as I hoped and intended and locked myself and Susan in the Tower. With my TARDIS.

I very much doubt if we shall meet again, and by the time you receive this letter Susan and I shall be long gone.

I remain, sir, your disobedient servant

The Doctor

The following notes were left over a number of years for Professor Chronotis at St Cedd's College, Cambridge. Rumours that Chronotis is actually a retired Time Lord have never been confirmed. The notes are reproduced by kind permission of the Master of St Cedd's.

October 2nd, 1955

Professor!

Called by and you're not answering your door. Hence this short note. I trust you're in good spirits? I'm doing fabulously well, thank you.

I was hoping for a cup of tea (milk, two lumps), but perhaps another time. Must dash – a bit of bother in the Horsehead Nebula that needs sorting out. But it's been such fun. We must do this again some time. Possibly some time when you're here too.

Yours in haste

The Doctor

April 11th, 1958

My Dear Chronotis

I called today, but was unlucky enough to find you out. The porter, Wilkin, kindly agreed to pass on this note and assures me you will be back after lectures at about 4pm. Assuming that's right, I hope to see you then. I really must tell you a rather enchanting anecdote I heard recently about a Venusian Shanghorn and a Perigosto Stick!

With all best wishes

The Doctor

July 24th, 1960

Professor – I imagine you already know, but I'm getting a degree!

An honorary degree in... Well, I'm not altogether sure what it's in, to be honest. A doctorate though – which certainly seems appropriate. And I'm sure I have you to thank for putting my name forward. How very kind – I am eternally grateful. But perhaps that's why you did it?

Anyway, I seem to have arrived a day or two early, but no doubt I shall see you at the ceremony. I hope I don't have to wear one of those silly hats. I've brought my own instead. And a scarf. Even though it's quite warm just now, isn't it?

Happy Times and Places

The Doctor

January 3rd, 1964

We wish you a merry Christmas and a happy new year!

Well, better late than never. Though obviously that's Christmas 1827. And the new year is 1217. Ha ha!

Sorry to miss you again, Professor. But I'm sure I'll be back sooner or later. Or earlier or later. It's so hard to predict, isn't it?

Hope the retirement thing is still going well. I really must try it some time. If the universe will give me some time off — possibly for good behaviour. Or more likely not.

Until we meet (or have already met) again

The Doctor

August 15th, 1977

Sorry — out of term. You're probably away somewhere.

The Eye of Orion is very nice this time of year.

D xxx

This letter was apparently discovered by Susan Foreman
tucked into her coat pocket after the Doctor left her on
Earth in the twenty-second century to marry David Campbell.
Susan and David had met during the Dalek occupation, both
working with the human resistance movement in London.
Both were instrumental in the attack on the Dalek mine in
Bedfordshire which led to the destruction of the Dalek task
force on Earth, and ended the occupation.

Reproduced by kind permission of the Campbell family.

My Dearest Granddaughter Susan

I cannot fail to notice the relationship that has been building up between you and young David. Much as I would like to, I feel I can no longer ignore it, or the possible implications. Whether you have yet realised this yourself or not, I am firmly of the opinion that the two of you are falling in love. I cannot in all honesty say that I approve. But David seems a fine enough young man, and it would be churlish of me to object and futile for me to try to put a stop to the thing. I know from experience that feelings and emotions cannot be suppressed or denied without the most unpleasant of consequences.

I hope to be able to tell you these things in person, but I write this letter in case I never find the right time. You would think travelling through the fourth dimension I would be better organised, but I am acutely aware that events may run away with us rather. So I would rather be safe than sorry.

Let me then say this: During all the years I've been taking care of you, you in return have been taking care of me. But that time – like all times – must come to an end. You are still my grandchild and always will be. But now, you are a woman too, and your life must change course.

I want you to belong somewhere, to have roots of your own, not meandering about the fourth and fifth dimensions in search of things we must accept we shall probably never find. So far from home – in space and in time. With David, you'll be able to find those roots and live normally like any woman should do.

I know that if I tell you this now, you will object. And I know from experience how stubborn you can be. But believe me, my dear, when I tell you that your future lies with David, and not with a silly old buffer like me.

One day, I shall come back. Yes, I promise, I shall come back. Until then, there must be no regrets, no tears, no anxieties. Just go forward in all your beliefs, and prove to me that I am not mistaken in mine.

Goodbye, Susan. Goodbye, my dear. Remember me.

Your ever affectionate Grandfather

This letter was originally written in 18th century French, the translated version here being filtered through the TARDIS Translation Circuits.

It was written by Jeanne-Antoinette Poisson - later Madame d'Etoiles before legally separating from her husband and becoming Marquise de Pompadour. The letter is dated shortly before her death in April 1764 at the age of 42.

The letter is signed 'Reinette' which means 'Little Queen'. This was a nickname she was given as a child after a fortune teller predicted that Jeanne-Antoinette would one day become mistress to a king.

The letter was entrusted to King Louis XV to deliver in person to the Doctor. Reinette had met the Doctor at various points in her life, and he saved her from death at the mechanical hands of mysterious clockwork automata...

My dear Doctor

The path has never seemed more slow, and yet I fear I am nearing its end.

Reason tells me that you and I are unlikely to meet again, but I think I shall not listen to reason. I have seen the world inside your head, and know that all things are possible.

Hurry though, my love. My days grow shorter now, and I am so very weak.

God speed, my lonely angel.

Suppressed for many years by the revolutionary French authorities, this letter was delivered to Marie-Antoinette as she awaited execution in 1793. The letter was suppressed for its obvious criticism of the revolution and the unrest and oppression that it predicts will follow. It is said, however, that Marie Antoinette herself found it comforting at a very difficult time – in declining health and facing execution.

It is not known exactly who delivered the letter to her, but it is addressed to her as 'Antonia' – as she was known as a child. She was born in Vienna in 1755, the daughter of Emperor Francis I and Empress Maria Theresa, and christened Maria Antonia Josepha Johanna. There is speculation that it may have been delivered to her by one James Stirling, an English aristocrat now known to have worked for the British government as a spy in post-revolutionary France under the pseudonym Lemaitre.

My dear Antonia

Forgive me for not writing sooner, but would you believe I have a job? Well, not a job in the usual sense, but my talents and expertise have been pressed into service.

I know that when this letter reaches you, it will be a difficult time. I can offer no hope, I'm afraid. France will degenerate following its revolution and a period will follow that is so harrowing it will come to be known as 'The Terror'. I know – I was there. Or in your terms, I will be there. I've seen first hand how your country will suffer.

But I know too that out of this suffering and misery will come a brighter future for France. And that is what I offer you now – hope. In your darkest hour, I trust that you will be able to cling to that and things will seem a little brighter.

It may also comfort you to know that I have not forgotten the happy time I spent with you and Louis, and I keep and treasure the exquisite pick-lock you were kind enough to give me.

I remain your friend

The Doctor

This short, handwritten message was left tucked between controls on the TARDIS console following the Doctor and Martha Jones's defeat of the Master and the Toclafane.

Martha did indeed need to call on the Doctor's services a while later. By then she was working for UNIT, and the Doctor returned to help deal with a Sontaran incursion that almost turned planet Earth into a Sontaran Clone World.

Doctor

I'm going to tell you this myself. I hope. That's the plan.

But in case I don't, here it is anyway. I might bottle out. I don't know, stood up to the Master, saved the world, but I'm not sure I can talk — I mean really <u>talk</u> — to you.

Because that's it, isn't it? It's taken all this for me to realise I'm actually good at what I do. I've spent so long with you, thinking I was second best but you know what? I'm good. And I've spent all this time training to be a doctor, so I can't waste it. I can't — it's not fair on the people I should be caring for and looking after.

Even if I tell you all that, I may not tell you this. It's like my friend Vicky, when we were students. She lived in the same house as this bloke called Sean and she fancied him something rotten. Loved him. Talked about him all the time. But never told him — and OK, he liked her. But that was it. She wasted years on him, and I doubt he ever even noticed. I told her she should get out. Over and over, I told her — 'Get out, and get your life back.'

So I guess what I'm saying is, it's time I took my own advice. It's time I got out. It's been great — not always fun, but I've had such experiences. And I'll take those with me, into my new life.

I'll leave you my phone. Either with this note, or I'll give it to you. If I need you, I'll call. I promise. So don't worry about me. Just get on with your life, same as I'm getting on with mine.

See you

Martha

This handwritten note was discovered among the possessions of Napoleon Bonaparte on Saint Helena following his death on 5 May 1821. It is unsigned, and undated.

Modern historians consider the letter to be a fake or a hoax as, despite being itemised in the list of Napoleon's possessions made in 1821, the note is written on pale blue writing paper watermarked 'Basildon Bond'. The Basildon Bond brand was not created by Millington & Sons until 1911 – some 90 years after Napoleon's death.

Thought I'd better write and say sorry. I'm still not really sure what I said to upset you, but I know you weren't happy last time I dropped in. The fact that you threw a bottle of wine at me was a bit of clue. Unless it was meant as a gift? Not a good way to deliver prezzies though - by air. Not glass ones, anyway.

So, sorry. There you go, I've said it.

I caught the bottle, by the way. Big thanks for that. Drank the wine with some friends on a picnic by a lake. They seemed to like it. I was disappointed, to be honest. Thought it would taste more like the gums. Funny how you forget things like taste when you change body.

Good job that didn't happen to your troops, I suppose. Imagine if they woke up one morning and didn't like the food suddenly. All that onion and garlic going to waste, or whatever you fed them on. 'An army marches on its stomach,' I told you once - remember that? You thought I meant they should crawl forward on all fours and I had to explain it.

Happy times and places. Still - the island's nice, isn't it? Good view. Better than Elba.

Martha Jones's medical training was not without incident. For one thing, she was involved in the bizarre mass hallucination incident when the entire staff of the Royal Hope Hospital, together with all the patients, believed that the building was for a time transported from London to the surface of the moon.

Following that, Martha was absent from her training for a sustained period, yet still managed to qualify as a doctor. Her student records included this handwritten explanation stapled to the page that noted her absence.

To Whom It May Concern

Note that Martha Jones has been absent from part of her medical training as she has been engaged on business of ~~planetary~~ national importance. I am not at liberty to divulge any further information on the matter as it is covered by the ~~Shadow Proclamation~~ Official Secrets Act.

If you need to know more, contact General Lethbridge-Stewart via the Ministry of Defence and tell him I sent you. Or if he's not available — which is quite likely as he's a busy man — then try Sir John Sudbury at Department C19. Unless he's resigned, which is quite possible.

Or if all else fails, then contact UNIT who are sponsoring Miss Jones's training. Or will be as soon as I can speak to them.

The Doctor   What doctor?
             Doctor Who?

Name:
M. Jones

Dept:
Medical Student

Authority: RH0776   Number: 0/75

Royal Hope Hospital

RH

The vault beneath Windsor Castle contains a plethora of assorted items. Some are antiques and artefacts owned by the Royal Family for hundreds of years. Others are gifts or trinkets that have been acquired more recently.

Although a catalogue is kept, the vault is usually in some disarray. It is therefore uncertain whether this handwritten note was obtained by the family and placed in the Vault, or whether it was left there by someone. It was found underneath a fez.

As I may be responsible for the worst miscalculation since life crawled out of the seas on this sad planet, I thought I should leave a note. If nothing else, you can take it as an apology for intruding.

I was actually looking for the Bow of Nemesis. But it seems it's no longer here. I've found the box and it's empty. Apart from a note saying that it disappeared in 1788 and that legend says that unless a place is kept for the Bow in Windsor Castle, the entire Nemesis statue will return to destroy the world. Well, you kept the box for it, but I'm afraid the statue's still returned.

Which is my fault.

So I really just wanted to say - if you have any idea where the Bow may have gone, please let me know. You can write on the back of this note. And if things go badly, I'll come back and check.

But hopefully it won't come to that.

Hopefully.

The two handwritten letters on this page are taken from the private archives of the late Harriet Jones, Member of Parliament for Flydale North and, for a while, Prime Minister of the United Kingdom.

Intriguingly, while both letters are signed 'The Doctor', they are in different handwriting. Archivists unfamiliar with the process of regeneration have therefore speculated that one or both of the letters may therefore have been dictated and handwritten by an unidentified third party.

It is clear that Harriet Jones was well acquainted with the Doctor, although he is not mentioned in her autobiography, *Do You Know Who I Am?*

Prime Minister - Fantastic!

Well, maybe not yet. But soon. Prime Minister. Really.

I know you'll do it. Of course I do. And I mean <u>know</u>, not guess or hope. Know. Once I'd remembered why I recognised your name. Nothing like being trapped inside the Cabinet Office in 10 Downing Street to concentrate and focus the mind, is there.

You did well. We all did. Hannibal and vinegar - who'd have thought, eh? Saving the world with pickled onions and a buffalo. Well, sort of.

You'll do great things. I know you will. And I mean <u>know</u>. A Golden Age, or as close as it ever really gets.

And if you need any help or advice, just give me a call. UNIT have a space-time telegraph system hidden away, but don't tell them I told you. They get a bit possessive.

See you

The Doctor

I so _so_ wish you hadn't done that.

But it's how the universe works, isn't it? Moments like that, decisions like that. For the best of reasons, I'm sure. You're not a bad person.

But I wish you'd thought a bit more.

I wrote to you before, and I told you – if you need any help or advice just ask. You should have asked. I think, after all I did on that day, I deserved that.

You'll be bitter, upset, angry. Of course you will. And quite right too.

But I was right, and one day you'll thank me for it. Well, you'll appreciate it. Well, you might understand. I _was_ right, by the way – you _do_ look tired. And I'm not surprised. Prime Minister – it's a big thing. I was President once, and I hardly lasted a day before I chucked it in. As far as I remember. It was several lifetimes ago.

So despite everything, I wanted you to know – you did well. Right up until the end. Usually, I reckon everyone deserves one second chance. Yours is that you finally get some time for _yourself_, some time to do what _you_ want to do, and not what you think you ought to do.

Use it well. Time is precious. I should know.

The Doctor

51

Found amongst the personal effects of General Sir Alastair Gordon Lethbridge-Stewart, this handwritten letter is undated and unsigned. It was one of the few documents that Sir Alastair insisted on taking with him into the nursing home where he spent his last few months.

On his death, it was found lying on the bedside cabinet, as if Sir Alastair had been reading it in the moments before he passed away.

My dear old friend... One of my oldest and dearest friends

I feel guilty that I haven't seen you for so long. Not for a couple of lifetimes probably. And now it's too late, at least for this lifetime.

But we've seen some times. Yeti, Cybermen, Autons and – of course – Daleks. I know you think I resented the time I was forced to spend with UNIT. But actually, looking back, it was among the happiest of times. They say that about being at school, don't they – you only appreciate it years later. I made such friends – Liz and Jo, Benton and Yates, Palmer and Bell. Harry Sullivan – who could forget Harry? Dear Sarah Jane, and of course – you.

I doubt I thanked you or showed much appreciation. I didn't really go in for that back them, did I? I've mellowed since. Or matured perhaps – even though I'm a younger man now than I used to be.

But that's enough about the 'used to be'. It's time to look forward, to the future. To the people we shall become.

I know that life gets harder as you get older, even if that means less to me than it must to you. But if it's any crumb of comfort, know that I am thinking of you.

I don't want to go, and I'm sure you don't either.

Splendid fellow – you always were.

The letter reproduced here was discovered by Matthew Puckering. It was tucked inside a paperback copy of Agatha Christie's novel *Death in the Clouds*, which he bought from a second-hand bookshop in the village of Lower Withering On The Spire.

The letter is undated, and there is no indication of whether it was ever delivered to Dame Agatha Christie. Given that the book itself was stamped 'Property of St Cedd's College Library', it seems unlikely.

ORIENT EXPRESS

Agatha

I've no idea if you'll ever get this letter – I've no idea if I'll bother to send it. Or be alive to send it. Or want to send it. But writing to you helps me organise my thoughts. You see, I've got a mystery to solve.

I'm on a train – the Orient Express. Well, sort of. It's probably no more like the Orient Express you know than it is the Blue Train. Except for the murders, obviously. Because it's flying. Through space. Really.

But that doesn't matter so much as the murders. Dead people. In mysterious circumstances. Got your attention now, haven't I? What happens is this – someone sees something horrible, and then 66 seconds later they're dead. No one else sees it, whatever it is. No one else dies, or not till next time it appears. Or doesn't appear. Bit of a poser.

This may be a red herring (why red, by the way? Why not a blue herring, or a yellow one?) but there is a legend of something called 'The Foretold'. It's said that if you see it, you die exactly 66 seconds later. Which can't be a coincidence.

But how does it all fit together? And what's it doing here on a train. In space? If you've got any ideas, I'd be really grateful. I mean, I saved you from a giant wasp, so I think you probably owe me.

If I ever deliver this letter.

Mind you, if I do, that means I survived, which means I worked it all out. So you may be off the hook anyway. Right, got to dash – Clara's managed to get herself locked in somewhere she shouldn't be. She's a lot of work, that girl. Don't tell her I said that.

Thanks for the advice, if you have any.

The Doctor

In the early years of the twenty-first century, the Mr Copper Foundation funded research into many scientifically advanced areas, including subwave networks. The Foundation was established and endowed by Mr Copper. His first name is not known, and it is widely believed that 'Copper' was in fact a pseudonym. There is no record of Mr Copper existing prior to Christmas 2007, and even after that date he seems to have no social security number or official records. Just lots of money.

This letter was found amongst other papers held by the Mr Copper Foundation. Its exact provenance is unknown, and experts believe it was probably written as a joke.

**Mrs Golightly's
Happy Travelling University
and Dry Cleaners**

Dear Mr Copper

Now that you've found your feet, and discovered just how far a million pounds will actually go, I thought it would be a good time to put you straight on a few things that Mrs Golightly might have missed off her Earthonomics degree course or got ever so slightly wrong. Having just spoken to Mrs Golightly herself, I think there may be quite a lot of them, so I suggest you do some research yourself and check the facts before — well, before you do <u>anything</u>, really.

Some of it you will have worked out by now anyway. So I'm sure you've realised that describing 'old London Town' as being in the country of Yookay is not quite accurate, any more than the notion that it is ruled by Good King Wenceslas. Yookay — or rather, the UK, is not part of Great Britain, and Great Britain is not separated from Great France and Great Germany by the British Channel.

As I dropped you off in time for Christmas, I'm hoping you now know that humans do not worship the great god Santa, who does not have fearsome claws or a wife called Mary. Nor does the Yookay/UK regularly go to war with the country of Turkey every Christmas Eve before eating the people of Turkey for dinner the next day. And the day after that, people do not start boxing each other.

Oh, and the space programme (such as it is) run by America (the country America, not someone called Ham Erica) until recently used a space <u>shuttle</u>. Not a space shuffle. That sounds more like a card trick.

Phonetic is not spelled with an 'f', although actually I admit that would make sense.

And a 'public convenience' actually refers to — oh, this will take forever. You'll find out. About everything. And you'll have such fun doing it.

I hope you get your proper house with a garden and a door, and its kitchen with chairs and windows and plates and everything. Above all, I hope you live a long and happy life. Planet Earth — what a terrific retirement plan! I'll try to drop in and see how you're doing soon.

The Doctor

ALTHOUGH HE IS A SELF-CONFESSED WANDERER AND CITIZEN OF THE UNIVERSE, THE DOCTOR DOES SEEM TO HAVE AN AFFINITY WITH CERTAIN PLACES. ONE LOCATION THAT SEEMS TO HAVE HELD SPECIAL SIGNIFICANCE FOR HIM IS COAL HILL SCHOOL, IN THE SHOREDITCH AREA OF LONDON.

Soon after leaving Gallifrey with his granddaughter Susan, the Doctor spent some time on Earth in 1963. Keen to learn something of the planet's history and culture, Susan enrolled as a pupil at Coal Hill School - presumably as it was the closest secondary school to where the TARDIS had landed, in a junk yard in Totters Lane. How long the Doctor was willing to spend in this one time and place is unknown, but Susan spent several months at the school, and was not happy to leave.

That said, Susan never really fitted in, which prompted two of her teachers - Ian Chesterton and Barbara Wright - to follow her home one evening. It was their discovery of the TARDIS that prompted the Doctor to leave - taking the two teachers with them.

It is now known that one thing the Doctor did during this period was hide the so-called Hand of Omega - a powerful stellar manipulator he had taken from Gallifrey. It was not until he was in his seventh incarnation that the Doctor returned to 1963 to retrieve it, only to find two opposing factions of Daleks had also traced the device...

Whether by chance, or perhaps due to the way she manipulated and travelled through the Doctor's own personal time line, Clara Oswald also became an English teacher at Coal Hill School. The Doctor too worked there alongside Clara and her friend, Maths teacher Danny Pink, for a while during the early twenty-first century, although not as a teacher. It is known that he did teach briefly at Deffry Vale High School, and in the guise of John Smith at Farringham School for Boys in Herefordshire in 1913. But at Coal Hill School his role as interim Caretaker was actually so that he could lure a dangerous Skovox Blitzer to the school and deal with it.

The Doctor's own academic record is far from impressive. It is recorded at Prydon Academy that he only managed to graduate with 51 per cent at his second attempt.

LETTER 29
An Application

From the First Doctor
Date: June 1963

The archives of Coal Hill School include this
application form and covering letter from 1963.
The school records indicate that Susan Foreman
was offered a place at the school and did indeed
attend for the majority of the Michaelmas term
(now usually referred to as the Autumn term).
The reason for her abrupt departure, which also
coincided with an unplanned sabbatical by two of
her teachers, is unknown.

76 Totters Lane
Shoreditch
London
Earth

21 June 1963 Anno Domini

To Whom It May Concern

Please find enclosed an antiquated paper form in application for a place at
your educational establishment for my granddaughter, Miss Susan Foreman.

I should warn you that both Susan and I have very high standards, and she
will not accept anything less than the most rigorous of academic tuition.

Rather than reply to the above address, I would be grateful if you would
hold any correspondence relating to this matter at the school office. I shall
collect it in person.

Yours faithfully

Doctor Foreman

COAL HILL
SECONDARY SCHOOL

APPLICATION FOR ADMISSION

FULL NAME OF APPLICANT: Susan Prydonian Foreman

SEX: Female

DATE OF BIRTH: 23 June ~~1866~~ by your calendar 1948

TERM FOR WHICH A PLACE IS REQUIRED: Michaelmas term 1963

AGE AT TIME OF ENTRY TO SCHOOL: ~~97~~ 15

FULL NAMES OF BOTH PARENTS (OR GUARDIAN): Doctor I. M. Foreman

ADDRESS: 76 Totters Lane, Shoreditch, London, Earth, Sol 3 Mutters Spiral.

TELEPHONE NUMBER (IF AVAILABLE): None - ironically

IF THE PUPIL WILL ENTER THE SCHOOL ABOVE THE FIRST YEAR, STATE REASON FOR APPLICATION:

Susan and I have been travelling and have only now settled in this time and place area. Against my better judgement, I have agreed that Susan may attend your establishment to learn something of the local customs and history as well as to acquaint herself with the level of learning and scientific expertise this planet has reached.

SIGNATURE: Doctor Foreman

DATE: 21 June 1963

FOR OFFICE USE ONLY:

APPLICATION RECEIVED: 19 June 1963

OUTCOME: ACCEPTED

APPEALED: N/A

Clara Oswald's application for a teaching position in the English Department is also held in the archives of Coal Hill School. The application and accompanying CV are of a standard format and largely unremarkable content. The file also includes the various references taken up by the school governors before deciding on the appointment of Miss Oswald. One letter in particular is worthy of note and is reproduced here.

Eastminster College, Whitaker Road , Eastminster, Surrey

Redundant. Of course it was for the duration of her study. When else would it have been?

12 March 2012

Dear Mr Coburn

Thank you for your letter of 3 March requesting a reference for Clara Oswald in support of her application for a post in the English Department at Coal Hill School.

How would I know?

~~As you know, I was~~ Miss Oswald's personal tutor here at Eastminster ~~for the duration of her study.~~ She was always an attentive student, and her diligence and perseverance are reflected in her excellent results.

Redundant. Again. Guessing you didn't actually qualify to teach English yourself then. PE more like.

~~In terms of character,~~ Miss Oswald has a positive attitude and an optimistic outlook. She sees the finer qualities in people and the potential in her students. Her teaching practice was exemplary, and she has a real affinity with children.

very round face

If I had to make one criticism, it would be that Miss Oswald does have ~~a stubborn streak.~~ *Tell me about it!* She is ~~a forceful character who is often determined to get her own way, and this can sometimes blinker her to the opinions of others and result in a lack of empathy.~~ *be of great help. No, really.*

I doubt it.

But this is a minor point, ~~and I am sure that in time and with experience she will soften and adapt. I~~ have no hesitation in recommending Miss Oswald for a position in your school.

Your sincerely

David Spooner

PS: Sorry for the handwritten annotations. My secretary obviously made some mistakes when typing up this letter, so I've corrected them for her. Didn't seem fair to make her type it all out again, or input the text or whatever you do in the early 21st century. But yes – Clara Oswald will be good. Give her the job. You won't regret it. Well, actually, you might. But give it to her anyway.

COAL HILL SECONDARY SCHOOL

HORAS

POLICE BOX

SHOREDI

63

A handwritten note from one John Smith, applying for the position of temporary Caretaker at Coal Hill School when the regular caretaker, Atif Basra, was taken ill.

Oddly, the letter is stamped as having been received on Monday 6th October 2014 although Mr Basra did not call in sick until Tuesday 7th. Obviously the date stamp is wrong.

SHOREDITCH

Chairman of the Governors : I. Chesterton
Headmaster : W. Coburn

Dear Mr Armitage

I am a close neighbour and friend of your Caretaker Atif. As he is sick with a sudden and unexpected bout of 'flu, I would like to offer my services as a temporary replacement until he is fully recovered.

I think you will find that I have all the requisite expertise and experience. Indeed, I was once informed by one of your predecessors that I was over-qualified for the post of Caretaker at Coal Hill School. But under the circumstances, I am willing to overlook my undoubted excess of skills and take the job anyway.

I have not bothered to attach a CV as you wouldn't have time to read it all. But in summary, I have worked in several schools previously (albeit on a temporary contract between other work) including Deffry Vale High School, Farringham School for Boys, and Prydon Academy.

In terms of the requisite skills for the job, I can assure you that I am well versed in cleaning, plumbing, electrical repairs, first aid, solid state micro-welding and many other relevant disciplines.

Obviously I am truthful (usually), tactful (sometimes), polite (probably), modest (despite my undoubted skills). I also get on well with children, and actually used to be one myself.

Please don't worry about replying to this letter. I shall report for work on Wednesday.

John 'The Doctor' Smith

Received on
October 6 2014

65

Danny Pink taught Maths at Coal Hill School. Prior to gaining a teaching qualification, he served in the British army, attaining the rank of sergeant. His service included a tour in Afghanistan, during which he saw front-line combat.

Following Danny Pink's tragic death in a road accident, this handwritten note was found in a drawer of his desk at Coal Hill School. Why it is a addressed to 'PE' is unclear. Possibly it was a nickname he had at the school.

PE

Can't you keep this place tidy? You have no idea how much work it is sweeping round all that scrunched up paper dropped on the floor. And I have to say that if you stacked the books on your desk in alphabetical order – or any logical order – it would make it a lot easier for me to see what you've been up to.

As for the tables, it looks like you've arranged this room as some sort of assault course. Which I suppose makes sense. But has it occurred to you even for a moment that having the desks arranged in regimented lines like troops on parade may not be the arrangement most conducive to the imparting of educational information? A classic 'horseshoe' shape might serve you better – try it. I'm sure you can organise a few ~~squaddies~~ kids into moving the desks round.

But it's up to you. Don't let me tell you how to do your job. Whatever it is.

Just so long as you let me do mine. So, sort the mess. And I don't want this happening again, or I shall have to talk to the head teacher about some form of detention.

Doctor Caretaker

SHOREDITCH
Chairman of the Governors : I. Chesterton
Headmaster : W. Coburn

Although this note is addressed to Mr Coburn, the head teacher of Coal Hill School, it was actually found among the personal possessions of Clara Oswald, one of the English teachers.

From the note's condition, it seems likely that Mr Coburn threw it away, having taken whatever - if any - action he deemed necessary. Miss Oswald, or someone else, then retrieved the note from the bin. Why she would want to keep it is unknown. Perhaps she found it amusing.

Dear Mr Coburn

I am writing to inform you that my short stint of employment at Coal Hill School is now coming to an end. I have achieved what I set out to do when I started here, and I am certain that Atif will soon be restored to full health and able to return to the post of Caretaker.

But it would be remiss of me not to point out a few things which I have noticed while I've been here.

First, it would be helpful to provide each classroom with its own supply of paper towels. These are essential equipment in the event of any 'spillage'.

Second, Mr Pink's classroom could be kept a lot tidier.

Third - actually there are so many I'm giving up on numbers now.

Next, the lunch rota could be organised far more efficiently. I have attached a colour-coded chart showing how each class can be fed and watered in half the time it currently takes. And while I think about it, the chips are a disgrace - they need to be much crispier if you want children to enjoy eating them. And may I suggest a selection of dips and dressings?

Playground duty seems to consist of a couple of teachers standing and bantering to each other outside the main hall, while the fights happen round the corner by the bins. Chart 2 (attached) shows a suggested patrol route for a team of two working in tandem. A look-out post erected at point 'A' on the chart would also help with all-round visibility.

Parents Evening. Although distracted, I did observe this ritual. It overran by a considerable length of time and many parents were kept waiting. A system of pre-booked (or allocated) appointments according to a rigorous schedule would help. To keep meetings focused, a bell should be rung at the start and end of every slot. At the bell, the teachers all stop talking immediately.

The Spark O'Clean Floor Polish is getting a bit low. I suggest placing a regular, repeating order.

I'll be sure to send you more thoughts as they occur to me, but I expect you'll want to take immediate action on all of these first.

Happy to help

John 'The Doctor' Smith (ex-Caretaker)

Assessing the Risk

Amended by the
Twelfth Doctor
Date: November 2014

In addition to the Risk Assessment held on file at Coal Hill School for the Year 8 Gifted and Talented Group Visit to the London Zoological Museum in October 2014, an annotated copy also exists.

There is no indication of who provided the annotations, or whether they were meant to be taken seriously. A sample page is reproduced here.

THE TIME LORD LE

Risk Assessment

And embarrassment – they're
Kids. Embarrassment is worse.

If one of them's PE
then that's not going
to be much help.

Area	Possible Hazards	Risk	Details	Control Measures
Museum Grounds	Slip, trip, or fall	Low	Risk of injury	Group supervised by sufficient adults at all times.
				One first-aid trained adult on hand at all times.
	Pupils leaving site while unsupervised	Low	Risk of children getting lost or being abducted.	Group always under supervision. Museum rear doors are alarmed.
Museum Galleries & Shop	Slip, trip or fall	Low	Risk of injury	Group supervised by sufficient adults at all times.
				One first-aid trained adult on hand at all times.
	Damage to displays or stock	Low	Financial risk to museum & implications for school	Group always under supervision by staff. Pupil conduct agreed (signed by parents/guardians).

So how come I found Maebh Arden wandering about on her own?

I'd be alarmed with all those Kids about.

Low? Are you joking? These are children you're talking about. It – Will – Happen.

Because money's all that matters, I suppose. Typical.

You just said all this. Why are you bothering to say it again? Do you think people won't read it?

Hang on – where do you cover the possibility of leaving the museum to find a dirty great forest has grown up overnight? What are you going to do about that? Oh someone might trip over – well, big deal, they can do that at home if they want. We're talking about a rapid global forestation crisis here.

Discovered amongst Charles Dickens's correspondence after his death in 1870, this undated letter appears to have been written on a page torn from one of the novelist's own notebooks.

In September 1860, Dickens burned almost all his personal correspondence, so it can be assumed that this letter was written after that. Experts suggest that the references to Dickens's little-known reading in Cardiff mean that the letter must have been written after Christmas 1869. However, the more contemporary style and vocabulary have led many to believe that it is a hoax, somehow inserted recently into the correspondence.

Dear Charlie

We never really got a chance to talk in Cardiff, did we - what with one thing and another. That's a shame because, like I said, I really am a big fan of yours. So I popped by for a chat, but you're not here. Probably off reading to some gathering somewhere.

I was at that reading you gave in February 1858, you know. Remember that? Seems like only yesterday. OK, for me it was only yesterday. But wasn't it amazing - £3,000 you raised for Great Ormond Street Hospital. In one reading. Fantastic! The reading was good too. I was hoping for a bit of Great Expectations, but as you hadn't written it yet that was never going to happen. Pity.

Still, keep bashing the words out. Keep the world entertained. I meant what I said about The Signalman, by the way. Terrific stuff. I know it must have been difficult, reliving that train crash. But I guess personal experience is a great inspiration.

And you'll have some other experiences - some other ghosts - to write about now.

See you

The Doctor

Christmas Absence

From the Eleventh Doctor
Date: December 2011

The disappearance of the Williams family in 2012 remains a mystery. Rory Williams and his wife Amy vanished from their house leaving no clue as to where they had gone. There were no signs of forced entry, or of foul play. Washing up was left in the sink, and shopping was on the kitchen table apparently ready to be unpacked and put away.

This paper napkin, with its cryptic note written in felt pen, was found pressed between the pages of a Melody Malone novel. (See also letter 113.)

I was here. Honest.

I arrived on time. Well, a bit early – you were out. Probably playing in the snow. Is there snow? I'll look. No, there isn't. Well, playing somewhere.

So yes, Christmas dinner. Here I am. I've brought my special straw to make the drinks more fizzy, a miniature sonic lance to carve the turkey, my own paper party hat, and everything.

Trouble is, I've got to shoot off again now. Bit of trouble in the Galpraneth Nebula, apparently. Something to do with a window cleaner, a slide trombone, and an exploding duck. I think that's what they said.

Anyway, I'll be back in ten minutes. Probably. I hope.

So I'm leaving this note in case I'm not.

In which case – sorry. Really, sorry. Again.

Paradise Towers was the last project worked on by the so-called Great Architect Kroagnon. A high-rise residential building complex, it won several awards for its design. However, Kroagnon came to believe that his designs would be compromised by people actually living in Paradise Towers, while he himself refused to leave. The matter was finally settled when Kroagnon was imprisoned in the basement.

When war broke out, those too old or too young to enlist were sent to Paradise Towers. But Kroagnon had programmed the automated cleaning robots to treat people as rubbish that needed removing. The society within the complex broke down into rival groups of young Kangs, the older Rezzies and the Caretakers who tried to keep order.

This letter was apparently slipped into the Doctor's pocket before he and Mel left Paradise Towers, having reconciled the Kangs, Rezzies and Caretakers - and having defeated Kroagnon, whose insane mind had possessed the Chief Caretaker.

PARADISE TOWERS

Dear Doctor

I know the Kangs have some sort of farewell ceremony planned, and I don't want to interfere with that. But we Caretakers can't let you go without passing on our appreciation for what you have done. The Chief Caretaker would have appreciated it just as much as we do, and it's a great shame he can't be with us now on account of him being unalive.

I do apologise for that business about the 3-2-7 Appendix 3 Subsection 9 death. All a misunderstanding, as I'm sure you are now aware. I can promise that should you return to Paradise Towers you will be greeted far more charitably. Possibly even with a 9-6-8 Paragraph 5 Subparagraph 7 Welcome.

But perhaps, now that the Rezzies and Kangs are cohabiting peacefully, it is time for us Caretakers to relax the rules. Maybe not everything should be By the Book. I'm tempted to say that this is a time for leniency and tolerance, for guidelines and advice rather than rules and laws. (There is provision for such relaxation in the Rule Book – Appendix 7 Subsection 3 Paragraphs 4 and 5.)

Yes, perhaps we should be a bit more accommodating of the old and the young, and not make them abide by the letter of the law. In fact, I think I'll make that a rule.

Thank you again Doctor – and if you and Mel ever return we'll keep an apartment ready for you, free of wallscrawl and with clean towels.

Build High for Happiness

Acting Interim Chief Caretaker

(Formerly Deputy Chief Caretaker)

Got to Dash...

From the Tenth Doctor
Date: 1599

In amongst the few surviving papers and correspondence of William Shakespeare, one short letter has puzzled and intrigued academics. It seems to have been written in 1599, or soon after. Although it makes reference to Hamlet, which is believed to have been written some time between 1599 and 1602, unfortunately it does not help us to determine the exact date the play was written.

Those familiar with the Doctor's adventures will appreciate that the letter was written soon after the Doctor and Martha Jones helped Shakespeare to banish the Carrionites into the Deep Darkness.

William

I am so sorry. I am so so sorry. I was hoping to stay and have a bit of a natter down the tavern. But then the Queen arrived and, well, things got a bit sticky as you saw. Not quite sure what her problem was with me, to be honest. But I expect I'll find out one day.

Anyway, just thought I'd leave you a quick note to say sorry for dashing off, and how brilliant it was to meet you. Martha was dead impressed. Thanks for your help with the Carrionites, by the way. Top stuff.

And feel free to use the words and phrases that I noticed interested you. Sycorax, and all that. Seems only fair. You've got so much still to write, so much to tell.

Just one thing, and this may sound a bit weird, plus I shouldn't give too much away, but when you come to write a play set in Denmark you may get a visit from another Doctor. Well, not really another Doctor, but, well, he'll be sort of me. Only different. But basically, give him the benefit of the doubt. It's all rather complicated and tricky to explain, even to a genius like you, because it gets sort of wibbly-wobbly and timey-wimey.

But anyway, here's the point — if you happen to have sprained your wrist writing sonnets or something, then don't feel bad about asking the Doctor (other Doctor) to write out the play for you. He'll love it.

Oh, and when he asks you for a pencil and you give him one, and he says "2B or not 2B, that is the question", please just laugh and pretend it makes sense and is terribly funny. He'll love that too. Strange bloke, though I say so myself.

Anyway, got to run, things to do. Apart from avoiding Queenie, the Face of Boe was in a spot of bother and I said I'd help him out. Didn't quite go to plan, I'm afraid. But that's another story.

See you soon. Though for me it was ages ago.

The Doctor

While investigating strange disappearances and power losses in Colchester, the Doctor found it necessary to work for a time 'under cover' in the Toy Department of Sanderson and Grainger. His letter of application is preserved in the Doctor's personnel file, still held by the company.

Other notes on the file relate to the difficulty of getting the Doctor to divulge his National Insurance number, or to provide a P45 from his previous employer. A tortuous and unhelpful series of letters to and from HM Revenue and Customs comes to an abrupt end because the Doctor simply failed to turn up for work one morning.

Care of Craig Owens
~~79a Aikman Road~~
~~Colchester~~

No, hang on – he's moved.
I'll find out the address and send it on later. No problem.
But it's in Colchester somewhere.
Nice little house. With a back garden.

Dear Mr or Mrs or Miss or Ms Sanderson and/or Grainger

I am writing to apply for a position in your lovely shop in Colchester. I don't really mind what I do, but it's very important that I work for you, if only for a few weeks.

Having visited the establishment, I think the Toy Department would benefit most from my undoubted talents. I love toys. I could play with them all day. And if I worked in the Toy Department, I promise that I would.

I am reliable (usually), smart (bow tie and everything), I get on well with people (unless they don't get on with me), and my motto is 'Here to Help' which I think is just what you need in a shop like yours.

If you need references, though I can't imagine why, I can provide recommendations and letters of thanks for my being 'here to help' from the ruling administrations of various star systems, President Nixon (oh no, hang on, I think he died), and various Earth-based security organisations. Just let me know which you want.

And I'll see you for work on Monday morning.

Yours very sincerely

~~The Doctor~~ John Smith (Dr.)

Obviously has a sense of humour – will probably be good with the kids.

THE DOCTOR
Here to help

This note was saved to The Library's central processor, CAL, and appeared within the virtual reality world created within the system in the form of a handwritten letter. River Song - or rather the virtual reconstruction of her derived from the memory and personality data saved in a neural relay - found the letter on the dressing table in her bedroom several days after arriving in the virtual world.

River

I'm sorry. It was the best I could do. I saved you. Except being saved isn't the same as being saved, is it? But I couldn't let you go. Whoever you really are. I have your diary, so I suppose I could find out. Take a peek, see a glimpse of my own future. Except that would be cheating. Like you said, spoilers. The future is secret for a reason, and it should remain a closed book — literally in this case.

But I hope you like your new world. I hope it's better than the alternative. But I couldn't let you die. Not after all we've been through — you already, and for me yet to happen.

I don't know who you really are, but I'm glad I've met you. I'm looking forward to meeting you again. I sense that it won't be easy, but I know it will never be dull.

The tricks Time plays on us, eh? I meet you for the very first time, and it's the very last time you will ever meet me. Serves me right for mucking about with it so much, I guess.

Look after the others — Amita, the two Daves, Miss Evangelista. Oh and Doctor Moon and Charlotte of course. And anyone else you find in there.

And remember — we did good. Everybody lived, more or less.

Virtually yours

The Doctor

Get Well Soon

From the First Doctor
Date: July 1966

In the mid 1960s, Sir Charles Summer was head of the Royal Scientific Club. He was responsible for the planned link-up of world computers to be controlled by WOTAN - the Will Operating Thought ANalogue computer created by Professor Brett and housed in London's Post Office Tower.

However, as is well documented, WOTAN became unstable, and created mobile computerised War Machines in an attempt to destroy humanity on the grounds that the world needed to evolve beyond Mankind. WOTAN was able to hypnotise people, including Brett, into helping it. Sensing that the Doctor could oppose its plans, the computer hypnotised the Doctor's friend Dodo (short for Dorothea) Chaplet. Luckily, the Doctor was able to break WOTAN's conditioning, putting Dodo into a deep sleep for 48 hours to recover.

This letter was left with Sir Charles Summer and delivered to Miss Chaplet when she woke. By then, the Doctor, with help from Sir Charles, had destroyed WOTAN.

The Royal Scientific Club

My dear Dodo

Forgive me for not saying goodbye. But by the time you wake up and read this, I hope I shall be on my way. Just so long as I can sort out this unpleasant business with WOTAN.

I'm sorry it's all been a bit traumatic for you, what with all that hypnotism business. But a few days' rest and I'm sure you'll be as right as rain. Sir Charles Summer has very kindly agreed that you can stay at his house in the country until you are all recovered. His wife will look after you — but just make sure you behave yourself. I know what mischief you can get up to.

I apologise for not staying to see you. But farewells are not really my sort of thing. And as for dashing off again, well, you know I hate to linger for long in one place or time. I did once, yes once — for Susan, my granddaughter. Dear Susan. You remind me of her a little, you know.

But you are back in your own time now, and your own place. Goodness knows when I would next be able to bring you home.

But we've had some interesting times — you, me and that young Steven. Remember the business with the Monoids? Or that cad the Toymaker? And then there was all the fun and games in the American West, and the planet where we left Steven to sort things out.

Goodbye, my dear. I hope travel has broadened your mind. And who knows, perhaps one day we shall meet again.

I wish you a speedy recovery

The Doctor

Although the fifth Emperor of Rome, Nero Claudius Caesar Augustus Germanicus, encountered the Doctor, he was never actually aware of the fact. Circumstances forced the Doctor to adopt the identity of a noted lyre player, Maximus Pettulian.

Unknown to the Doctor, Pettulian was planning to assassinate Nero. The Emperor took against the Doctor in any case, and planned to have him killed in the Arena. However, 'Pettulian' vanished before Nero could go through with his plans. This note was left in Pettulian's quarters and discovered some time after the Great Fire of Rome, which started that same night.

My dear Caesar Nero

I was sorry to have to leave in such a hurry after setting fire to the plans for your new city. But then I was, as you may have guessed from my comments, already fully aware of your plans for me. A farewell performance in the Arena, eh? In amongst the lions and the alligators and what have you. Torn to shreds for the amusement of yourself and your friends.

Well, under the circumstances, you'll forgive me for not staying.

Oh, and just so you know, I'm not actually Maximus Pettulian, the famous lyre player. To be honest, I can't play a note. What do you think of that? Hmm? And remember all that business over dinner where I told you that the music was so soft, so delicate, that only those with keen perceptive hearing would be able to distinguish the melodious charm of the music? Well, it was a lie. For all you

pretended it was wonderful music, I wasn't actually playing at all.

Not that you'd have known if I had been. I have to say that you, sir, are one of the least perceptive people I have ever met. I very much hope that we shan't meet again.

Good day to you

The Lyre Player formerly known as Maximus Pettulian

Bedtime Story

From the Eleventh Doctor
Date: October 2011

George was a scared little boy. Actually, he was an alien Tenza but even his parents Alex and Claire didn't know that, and believed he was their own child. George's fears of rejection, and of monsters under his bed, became real and it was up to the Doctor to help him accept that his foster parents loved him and would not send him away.

The Doctor left this short story for George, to remind him how special he is.

To George - I know you like bedtime stories, so I wrote this one especially for you!

There was once a boy who wasn't like other boys. He kept himself to himself and didn't really make friends because – well, because he wasn't like other boys.

At school, the other children joined in and answered the teachers' questions. They played together in the playground. They sat and talked together when they ate their dinner. But he didn't, because he wasn't like other boys.

Then one day, he noticed a girl sitting on her own at dinner while the other girls talked and ate together. Later, he saw her on her own in the playground while the other girls played together. And in class, he noticed that she never answered the teachers' questions like the other girls and boys did.

So the next day, he went up to the girl and he asked her why she was all alone, and why she didn't join in. 'Oh,' she said, 'it's because I'm not like other girls.' And after that they soon became good friends.

But over the next few days, a funny thing happened. They both started to notice that the other boys and girls didn't always join in or talk or answer. And one by one they asked them why. And one by one, the boys and the girls explained that they weren't like other boys and girls. And one by one they all realised that everyone is different and special, and just because you join in and make friends doesn't make you any the less special or different.

And the boy thought he should tell his Mum and Dad what he'd found out. They listened while he explained about the girl, and about all his new friends. 'You're right,' his Dad said, 'everyone is different and special. It's just that other people don't always realise that.'

'But everyone is the same in one way,' his Mum said. And when he asked her what way that was, she gave him a great big hug. 'Everyone is loved,' she said.

From the Doctor - with love

Welcome Home

From the Ninth Doctor
Date: 2012

Adam Mitchell was a computer genius who hacked into the Pentagon's systems when he was just 8 years old. After leaving university, he worked for the billionaire Henry Van Statten, analysing and evaluating alien technology and suggesting ways that Van Statten's Geocomtex company could use it to make a profit.

After Van Statten's sudden and mysterious disappearance in 2012, Adam Mitchell returned home to Manchester. He seems to have shunned company and become a virtual recluse, rarely leaving the house.

This note was left for Adam's parents, Sandra and Geoff Mitchell, beside the burned out remains of their answerphone.

Dear Mr & Mrs Mitchell

I've brought your son Adam home in the hope that maybe you can knock some sense into him. I've rescued him from certain death, and shown him sights that few other humans have seen. But he still wanders off and gets into trouble. Being a bit of a free spirit is one thing, but planning to change the future for your own gain isn't showing initiative, it's dangerous.

Being intelligent and academically clever obviously doesn't guarantee common sense, or any feeling of responsibility. I mean, he's only gone and got a great info-spike grafted into his brain. Don't believe me? Just click your fingers. Honestly - humans.

Oh, and you'll need a new answerphone.

The Doctor

The Moroks once controlled a vast space empire. But over time, it faded and the Moroks' influence dwindled. One last reminder of their former glories was a huge Space Museum established by the Moroks on the planet Xeros. But even this gradually became neglected and forgotten, until it was visited only by the planet's governor and the Moroks stationed there.

Following a successful Xeron rebellion and the subsequent liberation of the planet from the occupying Morok forces, the Museum was closed. Most of the exhibits were sold off to help revitalise the emerging Xeron economy. But this display card from the Museum shows that at least some exhibits were removed before the final closure.

TIME AND SPACE VISUALISER

This device, sadly not working, converted energy from light neutrons into electrical impulses.

According to Ven Der Haff's law, mass is absorbed by light. Therefore light has mass and energy. As the energy radiated by a light neutron is equal to the energy of the mass it has absorbed, anything that ever happens, anywhere in the universe, is recorded in light neutrons. The Time and Space Visualiser, correctly tuned to an historical event, converts those recordings into a form that can be displayed visually. In theory it could replay any event from history, the results being shown on the screen at the centre of the device.

This Time and Space Visualiser was captured by the glorious forces of the mighty Morok Empire after the capitulation of Grantanimus Five. It is not known whether the machine is a prototype that was never completed, or if it has simply developed a fault and ceased to function.

Personally, I doubt if it ever actually worked. It would take a genius to align the neutron amplifiers properly and connect the results to the realisation matrix. Fortunately, as I happen to be a genius, I rather think I can get it working. So rather than leave it gathering dust in this woefully neglected museum, I have removed the Time and Space Visualiser and decided to accept it as a gift in return for helping you Xerons get rid of those unpleasant Morok characters.

This letter was probably given to the Doctor after his Christmas day meal at the Travellers' Halt with Jackson Lake, his son Frederick, and the boy's newly appointed nursemaid, Rosita Farisi. It was found in the pub that evening. As the envelope had been opened, we can assume that the Doctor read the letter, perhaps leaving it behind inadvertently after the meal.

The Travellers' Halt, London

25th December 1851

My Dear Doctor

I know you don't care to linger once your adventuring is done. So I thank you heartily for tarrying long enough to join us for Christmas dinner at the Travellers' Halt. It was, I trust you will agree, a most enjoyable repast and the company quite convivial.

I sense in your heart a degree of sadness. A sadness that I have experienced myself, of course. Under the circumstances, I hope you will not think me presumptuous if I offer a small piece of advice. Obviously, whether you take that advice is your own decision. But there are times, I believe, when it does us good to leave the cares and troubles of our inner worlds behind. When it is advantageous to live in the moment and enjoy what life has to offer.

Too often, Doctor, you see the darkness and the transience of our world. Of all worlds. But I know you are aware of the potential and the beauty, of the wonder of the universe. Just now and again, let it in. Let it melt that hard heart. Don't always think of us mere mortals as merely mortal, but also as the friends that we should like to be. If you will let us.

We have shared so much, my friend. Not just the Christmas meal, not just the danger and excitement of defeating the Cybermen. I know you perhaps better than I think you sometimes know yourself. So I hope you will forgive me if I am impertinent. And I hope you will remember me, as I promise I shall remember you.

And believe me, Doctor, when I say this — if I had to be anyone else, anyone in the whole of time and space, then I am glad it was you.

Your friend *Jackson Lake*

Henry Gordon Jago was owner and manager of the Palace Theatre in London when the Doctor and Leela encountered Magnus Greel, a war criminal from the fifty-first century masquerading as the ancient Chinese god Weng-Chiang. Working closely with Jago and also with the eminent pathologist Professor George Litefoot, the Doctor and Leela eventually managed to track down Greel and put an end to his murderous villainy. Jago and Litefoot became firm friends during the course of their adventures with the Doctor, and indeed continued to work together as investigators of the strange and unusual.

This note was preserved by Litefoot together with other mementoes of the Greel incident and their subsequent adventures. It was left by Jago for the Doctor, together with a carpet bag of various components from Greel's futuristic devices and equipment. One of the items it contained was the trionic lattice key to Greel's time capsule. The Doctor's possession of this component proved invaluable in his defeat of the criminal.

The Palace Theatre
PROPRIETOR MR HENRY GORDON JAGO

My dear Doctor

Contained in this capacious carpet bag, which I discovered inadvertently in the cellar, is a collection of sundry items of baffling meaning.

The Professor and I are keeping observation on the theatre and shortly hope to report to you the whereabouts of the mysterious Weng-Chiang.

Your fellow detective H G J.

PS - May I reiterate once again what a privilege and a pleasure it is for both the Professor and myself to be working with you and Miss Leela on this devilish affair.

THE DALEKS AND OTHER MONSTERS

THE HISTORY OF THE TIME LORDS IS INEXTRICABLY LINKED WITH THE
HISTORY OF THE DALEKS - DEFINED BY THE LAST GREAT TIME WAR THAT
ALL BUT DESTROYED BOTH RACES. SIMILARLY, THE DOCTOR'S OWN LIVES AND
TIMES ARE BOUND UP WITH HIS CONTINUING BATTLE AGAINST OPPRESSION
AND INJUSTICE IN GENERAL, AND HIS CONFLICT WITH THE DALEKS IN
PARTICULAR.

Born out of a terrible thousand-year war, the Daleks were mutated
and genetically developed by their creator Davros into creatures of
pure hatred. Encased in their armoured shells, they are pitiless in
their pursuit of power and their loathing of all other life forms.

Since their first meeting in the Dalek city on Skaro, the Daleks
have represented everything that is anathema to the Doctor. It is
fitting that it was the Doctor the Time Lords chose to send to
try to prevent the creation of the Daleks. They had foreseen a
time when the Daleks might conquer all other races and become the
dominant creature in the universe. The Doctor was given a choice -
avert their creation, or affect their genetic development so they
would evolve into less aggressive creatures.

He failed. But the Doctor knew that while they would create havoc
and destruction for millions of years, out of the Daleks' evil
would come some good. Out of the Doctor's failure, according to
some analysts, came the Last Great Time War...

But by not destroying the Daleks, the Doctor also proved what sets
him apart from the Daleks and so many of his other foes. The Doctor
is constantly battling monsters and aliens that would oppress
other races and ruthlessly exterminate their enemies. In his own
words: 'There are some corners of the universe which have bred the
most terrible things. Things which act against everything that we
believe in. They must be fought.'

The monsters and creatures that the Doctor has fought are too many
to name. But they include the Cybermen, Sontarans, Ice Warriors,
Slitheen, Krillitanes... He has faced down Sutekh the last of the
Osirans, the Great Intelligence, and the Fisher King...

There are, inevitably, casualties in the Doctor's fight. But a
universe without the Doctor there to battle against the monsters
scarcely bears thinking about...

Towards the end of the Thousand Year War between the Thals and the Kaleds on the planet Skaro, the Kaled Council grew increasingly apprehensive about the research work being undertaken by their chief scientist Davros and his elite scientific research unit. The Kaled Council charged a senior Councillor - Mogran - with investigating the work and reporting back with recommendations.

The Doctor was a key witness at the subsequent inquiry, and a portion of his expert testimony is reproduced here.

The recommendations of the report were never acted upon, however. Rather than allow any interference with his work, Davros betrayed his own people to their enemies, the Thals. The Kaled city was destroyed in a rocket attack using a formula Davros provided to penetrate the protective dome that surrounded it. Davros then sent his prototype Daleks to wipe out the Thals.

THE KALED COUNCIL

I confirm that I make this statement of my own free will, at the request of Councillor Mogran.

Some of what I will divulge relates to events in the future – not only on this planet but also on others whose existence you don't even know of. But my knowledge is scientific fact. I do not expect you to understand this, but I ask you to accept it.

In relation to the work being carried out at the Bunker, you know already that Davros has changed the nature of this from the development of a weapon to end your war with the Thals, to the survival of the Kaled race. His research has led him to believe that the chemical, biological and nuclear weapons used in the first centuries of the war have already started to introduce genetic mutations into the Kaled race.

You call these mutations 'Mutos' and banish them into the wastelands. But Davros now knows there is no way to reverse the trend. So he devised experiments to establish the final mutational form of the Kaleds. Once he knew its form, he set about devising a travel machine for the creature he knew your race would become. In doing so, Davros has created a machine creature, a monster which will terrorise and destroy millions and millions of lives and lands throughout all eternity.

He has given this machine a name – a Dalek. It is a word new to you, but for a thousand generations it is a name that will bring fear and terror.

Now undoubtedly Davros has one of the finest scientific minds in existence, but he has a fanatical desire to perpetuate himself in his machine. He works without conscience, without soul, without pity, and thanks to the genetic changes he is making to the mutated Kaled creatures, his Daleks are equally devoid of these qualities.

This letter, together with a Medal of Commendation, was delivered to the Most Hallowed Halls of the Grand Order of Oberon.

The circumstances of Orcini's excommunication, as mentioned in the letter, are unclear. But his name was indeed added to the Record of Honour. It is likely that all mention of his excommunication was at the same time expunged from the Order's archives.

Orcini's Squire, Bostock, was also added to the Record of Honour, as a Knight Associate - a position bestowed upon him posthumously.

TRANQUIL REPOSE
A PAUSE, BUT NOT THE END

To the Council of Knight Commanders of the Grand Order of Oberon - greetings.

Together with this epistle, I return to you the Medal of Commendation you awarded to your Knight Orcini. It is with regret that I inform you of his death, and also the death of his Squire, Bostock.

I am aware that Orcici had been excommunicated from your Order, but I trust that once I have related the circumstances of his death, you will see fit to reinstate him posthumously and add his name to your Record of Honour.

Hoping for an honourable assassination with which to end his admittedly chequered career, Orcini undertook an assignment from Kara on the planet Necros to kill the so-called Great Healer. In fact, the Great Healer was Davros, creator of the Daleks - an honourable kill indeed. Particularly so as Davros was creating an new generation of Daleks, using genetic material from the Tranquil Repose funeral establishment.

But Kara betrayed Orcini. The communications equipment she provided contained a powerful bomb which would detonate when Orcini signalled that he had infiltrated the catacombs of Tranquil Repose. The explosion was intended to kill both Davros and Orcini.

TRANQUIL REPOSE
A PAUSE, BUT NOT THE END

Realising he had been betrayed, Orcini killed Kara - who had been captured by Davros. With Bostock dead, and having been wounded himself in a gun battle with Davros's new generation of Daleks, Orcini detonated the bomb to destroy the new Daleks. Sadly, Davros escaped the blast, having been captured by a Dalek taskforce despatched from Skaro.

To the end, Orcini showed courage, initiative, and a disregard for his own life and safety where the lives of others were at stake. He died proud to have been a Knight of the Grand Order of Oberon.

I ask that he be remembered and venerated as such.

The Doctor

Honorary Knight Commander, as decreed by
Grand Order Article 953/667

When the Doctor left Gallifrey, he took with him the Remote Stellar Manipulator that the Time Lord scientist Omega used to detonate a star and create the energy needed for time travel. The Doctor hid the device, nicknamed the 'Hand of Omega', in a coffin.

When he returned for the device, in his seventh incarnation, he found that two rival factions of Daleks were fighting to obtain the Hand of Omega. Typically, the Doctor had a plan to destroy the Daleks - but that plan depended on the right faction obtaining the device.

When the Hand of Omega fell into the 'grubby little protuberances' of the wrong group, the Doctor sabotaged their Time Controller - the equipment they had used to travel to London in 1963 - and left this note written on a calling card marked with a distinctive question mark.

Yes, it's me.

Last chance - if you leave the 'Hand of Omega, and go now, I won't destroy you all.

Or stay, and face the consequences.

The Cabinet War Rooms were built beneath the Treasury building in Whitehall in 1938. They became operational in late August 1939, just a week before the outbreak of the Second World War.

Taking over from Neville Chamberlain in 1940, Prime Minister Winston Churchill conducted the war from these hidden, secret rooms - including overseeing the aborted Ironside Project led by Professor Edwin Bracewell. Although initially showing much promise, the project came to an abrupt end after the intervention of the Doctor.

This letter was found in the Filing Room several days after the Doctor's departure. It was filed under 'D' in a folder marked 'For the Prime Minister's Eyes Only'.

TOP SE

Winston

I meant it, you know. The world doesn't need me right now - it needs Winston Spencer Churchill. It needs you. Cometh the hour, and all that.

Oh, I can do all sorts of things with my TARDIS. But you know what? It's never enough. And it would be dangerous to do too much.

Yes, there are ways I could shorten the war. I could help you save countless lives. But History doesn't work like that. Sorry. And even if it did, it isn't just what happens now that's important - it's the effect it has on future generations. It's how people will look back on these dark days and say (if you'll forgive me stealing your lines) 'this was their finest hour'.

One day though, one day, I'll let you inside my amazing blue box. One day I'll give you a glimpse of other times and other worlds. I owe you that, my friend.

But for the moment, you have other work to do. It may seem tough. It may seem as if this is not actually your finest hour but your darkest night. But you can do it. I know you can. You know you can.

K.B.O.

The Doctor

TOP SECRET....TOP SECRET

RATTIGAN ACADEMY

You will by now have heard of the tragic death of Luke Rattigan. Coming so soon after the global failure of ATMOS, this is a double blow for the Rattigan Academy, of which I am sure you are proud to be a member.

Luke, as I'm sure you know, was a brilliant and talented young man. He may have been opinionated and arrogant - well, rude, actually - but this masked a deeper feeling of insecurity. I only met him briefly, but even so I got the impression that he was a man who felt unappreciated for much of his life. He looked at the world and he connected things - apparently random things. He thought the rest of the world was stupid not to make those same connections, not to see what he saw.

Whether he realised it or not, he founded the Rattigan Academy to help ensure that others as gifted as he was should not suffer in the same way.

And while, yes, I found him to be an unpleasant little oik, Luke Rattigan did finally come to terms with the world and his place in it. For all his intelligence and ability, it was only at the end of his life that he finally worked out how to be clever.

Another clever thing he did was to endow the Rattigan Academy as a trust - the Rattigan Trust - to which he left a large proportion of his substantial fortune. Which means that despite the loss of its founder and mentor, the Rattigan Academy will be happy to welcome you back to continue your research, backed by a generous grant. The Trustees will be in touch shortly to discuss the details with you.

As he was dying, Victoria Waterfield's father, the scientist Edward Waterfield, asked the Doctor to look after his daughter. Waterfield was killed when he threw himself in front of the Doctor as a Dalek fired on them both.

A quiet and reserved young woman from the mid nineteenth century, Victoria took a while to adapt to the rather hectic life on board the TARDIS. But she became firm friends with both the Doctor and his other travelling companion, Jamie McCrimmon - a young Scot the Doctor rescued from the mid eighteenth century.

Eventually, however, the strain and anxiety of the Doctor's adventures came to tell on Victoria and after a particularly harrowing encounter with a homicidal seaweed creature in the 1960s, Victoria made the decision to leave the Doctor and settle down. Frank and Maggie Harris, who had both been involved in battling the Seaweed Monster, offered to look after Victoria.

This letter was posted the evening before the Doctor and Jamie left in the TARDIS, and delivered the day after they had gone.

ESGO

Victoria, my dear

There are some decisions that we have to make on our own. All the advice in the world shouldn't make any difference to them, because they are _our_ decisions. They affect _us_. So, I wouldn't dream of offering you advice. But for what it's worth, I believe you are making the right choice.

If you stayed with us, I could never promise that Jamie and I could get you home. And anyway, what sort of home would you have. We've spoken before about how we miss our families, but I should tell you that I owe your father a lot. I owe him my life, and just as important I owe him the time that I have spent with you.

I know it hasn't always been easy. Jamie can be quite tiresome sometimes, can't he? I never promised you safety, but I think – I hope – you have enjoyed your time with us. The places we have been, the times we have visited, the people we have met. There's been danger, but there's been excitement too. Our lives were so different to everyone else's – no one in the universe could have done the things you have done. Travel really does broaden the mind, you know.

I shall miss you, Victoria. I hope that you will miss me too. But times change, as I know better than most. It is time for you to start a new chapter in the book of your life and, with Maggie and Frank Harris to look after you, I am sure that the chapter will be a good one.

Perhaps we shall meet again. I hope so. But until that time, be assured that you will sleep in my memory, just as I hope that I shall sleep in yours.

Your friend

The Doctor

Dr Henry Black is one of the foremost experts on the nineteenth-century painter Vincent van Gogh, and has written widely on the artist's life and work. His introduction to the catalogue of the 2010 exhibition of the artist's work at the Musée d'Orsay in Paris gives a concise introduction to the painter.

Several years after the exhibition, Black was surprised to find that his own copy of the catalogue had been annotated at some point following the exhibition.

BORN on 30 March 1853 in Zundert in the Netherlands, Vincent Willem van Gogh is arguably the greatest artist who ever lived.

Yes, I'd argue for that too

But his early life gave no hint of the greatness that was to come. He worked for a firm of art dealers, travelling between The Hague, Paris and London, and for a while toyed with the idea of becoming a pastor. Indeed, for some time he worked as a missionary in Belgium. It was then that he began to sketch the local people.

The Potato Eaters? What sort of title is that for a great painting? They'd prefer chips anyway.

Although he drew as a child, he did not start painting until he was in his late twenties – relatively late in his tragically short life. His first major work, *The Potato Eaters*, was painted in 1885. Moving to Paris in 1886, he discovered the French Impressionists, and was greatly influenced by their work. Another influence was the strong sunlight he found when he moved from Paris to the south of France. From this he developed his distinctive and immediately recognisable style of painting.

It wasn't just his style. That was how he saw the world – actually how he saw it. Through his eyes. In a different way to you and me.

Bit weird. But useful if there's a stray Krafayis about.

Once he got started, though, his output was prodigious. He produced over two thousand paintings (oil and watercolour), sketches and drawings in the last ten years of his life. His best known and most accomplished works were produced in his final two years. This was when he painted the *Portrait of Dr Gachet*,

Really? They'd take up a lot of space. He must have had a garage or something I never noticed.

THE TIME LORD LETTERS

VINCENT WILLEM VAN GOGH

The Starry Night, Wheatfield Under Thunderclouds… Who knows what else he would have achieved had he lived longer.

> What about The Church at Auvers. I mean, that's probably the most important painting of them all, given what it originally showed.

Throughout his life, Van Gogh suffered from anxiety attacks and mental illness. While this probably informed his art, it also left him frustrated when it meant he could not work. In Auvers-sur-Oise, on 29 July 1890, his illness finally got the better of him and Van Gogh took his own life. It is widely accepted that he shot himself, dying from the wound, although no gun was ever found… He was just 37 years old. *So sad…*

> Tell me about it.

But to me, despite his relatively short career, Van Gogh is the finest artist of them all – the most popular and beloved of the great painters. His command of colour is magnificent. He transformed the pain of his tormented life into ecstatic beauty. Pain is easy to portray, but Van Gogh used his passion and pain to portray the ecstasy and joy and magnificence of our world. No one had ever done it before. Perhaps no one ever will again. To my mind, that strange, wild man who roamed the fields of Provence was not only the world's greatest artist, but also one of the greatest men who ever lived.

> You should see his Exploding TARDIS. Well, it's my TARDIS actually, but you know what I mean.

Henry Black

What exactly happened on Storm Mine 4 has never been officially disclosed. The files on the incident remain sealed in the Government Archives in Kaldor City. Some say it was an accident, but there are also rumours that the events that led to the death of all but three of the sandminer's crew amounted to nothing short of a robot revolution. The rumours are uncomfortably close to the truth...

What was disclosed was that Commander Uvanov and Pilot Toos survived the incident with only minor injuries. Chief Mover Poul also survived, but suffered severe after-effects, having developed Grimwade's Syndrome. Commonly known as Robophobia, this is an unreasoning dread of robots, possibly brought on by the fact that while the most advanced Dum, Voc and Super-Voc class robots are humanoid, they give off no human non-verbal signals or feedback. This lack of body language undermines certain personality types causing identity crisis, paranoia, and in extreme cases - as with Poul - personality disintegration.

The truth may never be made public, but this handwritten note was left in the main crew room on Storm Mine 4.

Dear Uvanov and Toos

I am so sorry Leela and I had to rush off before you woke up. But I'm sure you'll soon be feeling just fine. There's a rescue ship on the way, and I rather suspect the crew will want some sort of explanation about what's been happening on this Sandminer. You'll be so much better at explaining than I would. We're not even supposed to be here – which is a bit awkward and means even more explaining. And then there's all the not being believed, getting locked up, having to escape and prove what I said was right all along. Tedious.

So I think it's best if we just go on our way and leave it with you.

Look after Poul. He should be fine, too, though I doubt if he'll want to be anywhere near a robot for a while. Maybe you won't either, but in Poul's case the condition is rather more severe. Remember to warn the rescue crew that he has quite advanced Grimwade's Syndrome. Definitely no robot nurse for him.

As to what else you tell them, well that's up to you. But maybe it would be best if they think there was some sort of accident or malfunction rather than a robot rebellion organised by a mad scientist who rather wanted to be a robot himself. That could cause some pretty extensive consequences that I'm not sure your civilisation is quite ready for yet.

Anyway, must dash. I've promised to take Leela to the theatre, and I'm hoping to catch Little Titch.

Happy Times and Places

The Doctor

An unexpected and apparently motiveless attack by the Sontarans on Space Station Camera, a highly regarded research facility, left the entire crew dead. Following the incident, a Database of Condolences was set up at the nearby Magnostellar Institute.

Reproduced here is one record from that database.

Name of Deceased: Dastari, Joinson

Position on Space Station Camera: Head of Projects

Record Created By: Smith, John (Dr)

Relationship to the Deceased: Friend / Colleague

Joinson Dastari was one of the cleverest and most qualified people I know — with enough letters after his name for two full alphabets. A pioneer of genetic engineering, if he had a fault it was that he sometimes let the wonder of discovery and invention cloud his moral judgement. Just because something can be invented, does not necessarily mean that it should be. I have no doubt, for example, that he could have genetically augmented an earwig to the point where it understood nuclear physics. But it would still be a very stupid thing to do.

So, I admit, I took issue with him over his experiments to augment an Androgum. After nine operations, he augmented Chessene O' the Franzine Grig to mega-genius level. Again, I would submit, this was clever, but also very stupid.

Of course, the ultimate ends of his work will never be known, as that work died with him.

I well remember attending the inauguration ceremony for Space Station Camera, and seeing Dastari appointed as Head of Projects. Dastari was just one of the many great scientists and the lesser known but no less loved and valuable technicians, assistants and service personnel to die in the unprovoked and brutal attack on Space Station Camera by the Ninth Sontaran Battle Group under the command of the late, but unlamented, Group Marshal Stike. The contributions to science of Kartz and Reimer, Haylman and Scwartzel — and of course of Joinson Dastari — will never be forgotten.

But of all of them, Dastari was my friend, and I shall miss him.

For almost 150 years, the town of Mercy in Nevada has kept itself isolated from the rest of the world. With a population that has rarely risen above a hundred, it has resisted any form of expansion. Its location in the middle of the desert has ensured that it has not become an attractive or convenient location. Transport links are limited to a single, infrequently used highway, and Mercy seems to exist only to serve the local farming community.

The town does, however, have a small museum. This is located in the former Marshal's Office. The intriguing items held by the museum include components of an early electric lighting system, a journal dating back to 1870 said to have been kept by one Carl Jecks, and the document reproduced here.

I, the Doctor, by the power vested in me by the people of the town of Mercy in the state of Nevada do hereby relinquish the position of Marshal.

I was appointed Marshal on the death of the previous incumbent, Isaac. (Sorry, didn't get his surname. Maybe it was Marshal, that'd be funny wouldn't it? Would make him Marshal Marshal. Right, moving on...) He was a good and fair man, and I hope I lived up to his high standards and the equally high expectations of you good people of Mercy.

In my place, I appoint Kahler-Tek (also known as 'the Gunslinger') as the new Marshal of Mercy. Like me, he isn't exactly local, but he'll do a good and conscientious job. A job he was made for. Literally.

But by now, you must be used to the strange and the impossible here in Mercy. So Kahler-Tek will fit right in. Where he comes from doesn't matter. The fact he won't use the Marshal's office, or even venture into town much, doesn't matter either.

And if anyone new to the town asks why you don't have a Marshal or a Sheriff, just tell them you've got your own arrangement. You needn't tell them what it is. You needn't tell them you've got your own special lawman watching out for you. Your very own lawman who fell from the stars.

Ain't no one gonna mess with him, as you might say round these here parts. Yes. Better stop now, getting carried away.

So, in summary: Doctor and deputies Rory and Amy resigning. New man Kahler-Tek is now Marshal.

Ex-Marshal Doctor

Elton Pope, former Transport Manager, was caught on the periphery of several alien invasions. He was out shopping when the Nestene Consciousness activated its killer Auton shop dummies. He heard the Slitheen ship crash into the Clock Tower of the Palace of Westminster, and he witnessed the arrival of the Sycorax.

But his first encounter with the Doctor was when he was just three or four years old. He glimpsed the Doctor in his house in the middle of the night – the same night as his mother died. In later life, the incident haunted Elton, and he joined a group dedicated to finding out more about, and possibly tracking down, the mysterious Doctor.

The other members of the group subsequently vanished. One of the mementoes that Elton retained from that time is this letter.

Dear Elton

I am so very sorry.

I should have come and found you years ago. I meant to, I really did. But time goes by so fast. And now you've come and found me. Well, I can't blame you for that. Probably would have done the same myself.

And it meant you met Ursula. Don't feel bad about what happened to her – and to the others in your nice little gang. It would have happened whether you were involved or not. Blame the Abzorbaloff, blame Victor Kennedy. Don't ever blame yourself.

As for when we first met all those years ago... Well, it wasn't really a meeting, was it? You, a frightened and confused little boy watching me in your living room. The night your mum died. I stopped the living shadow that hid in your house when it escaped from the Howling Halls. I stopped it from getting to you. But I was too late to save your mum. I can't save everyone, I only wish I could. If I have one regret, it's that I can't save everyone, no matter how hard I try.

You just have to do your best.

I saw your video – neat job, by the way. And you're right you know, when you say that when you're a kid, they tell you it's all grow up, get a job, get married, get a house, have children, and that's it. You're right – it's not. That's not "it" at all. Like you said, the world – the universe – is so much stranger than that. It's so much darker, and so much madder. And so much better.

Elton Pope – philosopher. Who'd have thought?

Do your best **The Doctor**

L.I.N.D.A.
London Investigation 'n'
Detective Agency?

Beyond the year 200,000, the Game Station broadcast over ten thousand entertainment shows to Earth. Run by the Bad Wolf Corporation, the Station's shows included many that were lethal, including *Call My Bluff* (with real guns), *Countdown* (with 30 seconds to stop the bomb), *Ground Force* (contestants turned into compost), *Stars in Their Eyes* (sing or be blinded) and the deadly *Wipeout*.

Removed from the TARDIS by the Controller of the Game Station, Captain Jack Harkness found himself threatened by robots TRINE-E and ZU-ZANA in an edition of *What Not to Wear*, while Rose Tyler faced the deadly Anne Droid in *The Weakest Link*.

The Doctor, however, was transported into the *Big Brother* house – where eviction meant death. This is a transcript of what he said during his first visit to the Diary Room, recorded soon after he arrived on the Game Station.

You have got to be kidding.

I was in the TARDIS. With Rose and Jack. I bet one of them pressed something. I bet I know who and all. I told him, 'Don't fiddle.' I told him. It's like when I tell him, 'Don't flirt' - red rag to a bull.

So, we left Raxacoricofallapatorius, then whizzed off to Kyoto. Japan, 1336. Nice little break, I thought. Well, I was wrong about that. Bit of trouble with an escaped Rawshark from the Helenouin Drift. Sorted that, and the locals came after us. Said we were responsible for all the deaths and the damage to the fishing boats. Never happy, humans. We were lucky to escape.

Then, back in the TARDIS, all together, laughing. And then there was this light. This white light coming through the walls. And here I am. So is that your doing? Because I'm not staying here. I'm getting out, I tell you now. I'm going to find Rose, and I'm going to find Jack. And then I'm going to find you. And when I do, I want some answers.

Because this isn't some game, is it? Not with a transmat that can pick me up from inside the TARDIS. It'd have to be about fifteen million times more powerful than your bog-standard transmat. I don't know what's going on here, but I'm going to find out.

Fair warning.

Following the apparent death of Klineman Halpen, Chief Executive of Ood Operations, at the company's main facility on the Ood-Sphere, the company's Directors were summoned to an extraordinary board meeting. This letter was sent out to all members of the company's board ahead of that meeting.

The truth about the company's operations to which the letter refers was that Halpen (and his predecessors) were genetically altering the Ood. By removing the creatures' hind brains, they made them docile and servile. Halpen was also suppressing the influence of another, huge collective Ood brain, buried beneath the surface of their planet. In effect, the Halpen family, through their company, had enslaved a peaceful and intelligent race.

You don't know who I am. But I know who you are. A copy of this letter is going out to all of you - all the members of the Governing Board of Ood Operations. And it's important, so read it.

I guess you all know by now that Klineman Halpen is dead. Well, he isn't.

And you probably don't know half the truth behind the success of Ood Operations. From the paperwork, it looks like he kept you in the dark, like he kept everyone else in the dark. I hope he did. If he didn't, if you knew, then I just hope you can sleep at night. Actually, that's a lie - I hope you can't.

Attached to this letter is a report by the late Dr Ryder, which he compiled over several years of working closely with Halpen. It shows how your company has exploited the Ood. How your company has mutilated and subdued them, turning the creatures into slaves.

You have a board meeting next week to decide how the company can proceed without a Halpen at the helm for the first time since it was established. Like I said, Klineman Halpen isn't dead. But he won't be running the company. Because he's now an Ood himself, mutated by a regular dosage of Ood-graft.

If he was in a fit state to advise you, under the circumstances I think he'd tell you the same as I'm going to tell you. Which is this:

Stop it. Stop it now. Close the company with immediate effect. Walk away with whatever is left of your consciences intact while you can. Leave the Ood-Sphere, and let the Ood continue their lives, free at last.

You have one chance to put things right. One last chance. I won't give you another.

The Doctor

Preserved in the Cranleigh Family Archive, this note was left on the drinks cabinet in the drawing room of Cranleigh Hall following the funeral of George Cranleigh, the noted explorer. Believed lost in the Amazon the previous year, Cranleigh returned to the family home in June 1925, only to die in a fall from the roof while apparently trying to escape a fire that had broken out in a wing of the house.

Cranleigh is best remembered for his discovery and cataloguing of rare plants. His book *Black Orchid* documents his search for - and discovery of - one of the most sought-after specimens. The only black orchid known to exist outside the Amazon basin is still kept at Cranleigh Hall.

My Dear Lord Charles and Lady Cranleigh

I felt I should write to offer you both my thanks and my condolences.

I am deeply sorry for your loss. It cannot have been easy caring for poor George these past months. But, under the circumstances, perhaps his death is a merciful relief. Nevertheless, I am sure you are deeply saddened by his passing.

On a happier note, I must thank you, both for myself and on behalf of Adric, Nyssa and Tegan, for the hospitality you have shown us. It's a long time since I played cricket, which is one of my most favourite pastimes. So I was delighted to be able to indulge in it again.

Thank you too for the gift of George's book – 'Black Orchid'. I shall treasure it, just as I am sure you will treasure his memory.

With all best wishes

The Doctor

Following his death from pleurisy in August 1848, George Stephenson's third wife Ellen made an effort to sort and catalogue his papers and correspondence. Much of it was discarded, but this undated handwritten note was retained as a curiosity.

The letter was obtained by agents of the Torchwood Institute in the late nineteenth century, and now resides in their Central Archive.

George

All those years ago, when you were in Killingworth, do you remember what I told you? 'Your invention will take off like a rocket,' I said. And I was right, wasn't I? (Rhetorical question – I'm always right.)

I confess, I did have something of an advantage. I knew what would happen. I knew all about the Stockton and Darlington Railway, and about the Liverpool and Manchester too. Even the skew bridges – though I have to say I'm a bit less convinced by those.

I meant to write sooner, and apologise for the way Peri and I just rushed off that day, after sorting out the trouble with the miners and their lack of sleep. But we had places to go and times to see. I did intend to pop back and see you and Lord Ravensworth. But somehow it never happened.

Steam engines, though! I've always wanted to drive a steam engine. So who knows, maybe one day I'll turn up on your doorstep with my engine-driver's hat ready for a brisk outing. Full steam ahead! Maybe I already have.

In which case, I'll see you earlier.

The Doctor

PS – Sorry about the paper – I'm currently trapped in a Woolworths by a group of rogue Quarks and this is all I could find.

Chloe Webber is one of the most talented and acclaimed artists to have emerged in Britain in the past few years. Still in her teens, she has exhibited at galleries across the country. Her surreal paintings have an almost childlike quality, perhaps not surprisingly, given her age. They are noted for the recurring motif of a blue box, which appears somewhere in every picture.

The point of the blue box is unknown, and the artist has always declined to comment on its inclusion or meaning. But it may have been inspired by this letter and drawing, which are reproduced here for the first time with the kind permission of Miss Webber.

Dear Chloe

Sorry I had to rush off — small job at the Olympic Stadium. Involved some running and carrying a torch. You'll probably see it on the telly.

More importantly, it involved sending the Isolus child back to its family, back to its brothers and sisters. The Isolus pod is powered by the heat and hope, the courage and love of the Olympic Flame. It's free again now, and so are you.

I know things worked out for you at home. Rose told me all about the missing children coming home, and that you finally conquered your fear of your father. It's been a difficult time for you, I'm sure. But things will be all the better for it in the long run — and I should know, I've just done a long run.

Oh, you'll be frightened again, we all are. But just sing to yourself. Give that kookaburra in his old gum tree some wellie. Remember that things will get better. Keep drawing pictures, but pictures full of promise and joy.

And don't forget us, will you? To help you remember, here's a picture I drew. A bit boring, I know — just a blue box. But it's something that means a lot to me. It's what I think about when I get frightened. It means safety and security and friendship.

Say hello to your Mum for me. Maybe one day I'll come back and see how you're both doing. Although I already know — you'll be doing brilliantly.

The Doctor.

For centuries the people of Tigella lived in an underground city, away from the aggressive vegetation that covered the planet's surface. The power for the city - its light, heat and ventilation - came from a mystical object known as the Dodecahedron. It was said that the Dodecahedron had descended from the heavens as a gift from the great god Ti.

Quite what happened to the Dodecahedron is unclear, but it is known that one day it disappeared from Tigella as mysteriously as it had arrived. The Doctor's exact role in events is also rather vague. But he did leave this letter for the Tigellan leader Zastor shortly after the Dodecahedron vanished.

A new beginning – how exciting!

I wish I could stay and help with the gardening. But I'm sure you'll manage to get those bell plants under control, and Romana and K-9 and I have things to do. Got to drop our friend from Earth back home for his tea for one thing. And Romana tells me we're needed back on Gallifrey. Though between you and me, Zastor, I doubt they need us at all. Or no more than usual, anyway.

Life on Tigella will be very different without the Dodecahedron. As well as clearing away some of the lush, aggressive vegetation so you can move out of that underground city, you'll still need some source of power. Sunlight is good – and not just for the plants. Or hydro-power. Or the energising of hydrogen...

But that's all for you to sort out. As I said, an exciting new beginning.

And perhaps in another fifty years I'll come back and see how you're getting on. Perhaps I'll be ready for a spot of gardening myself by then. So long as you're not cultivating Krynoids.

Have such fun, won't you?

The Doctor

THE TIME LORD LETTERS

A favourite of conspiracy theorists in the early twenty-first century was the so-called Easter Egg Man. Hidden on seventeen different DVDs, apparently at random, is footage of a man speaking directly to camera. The supposition was that the footage showed one side of a conversation, and sparked speculation as to who the man might be talking to as well as what he might be talking about. Phrases such as 'Don't blink' and 'The Angels have the Phone Box' became popular on social media, conspiracy forums, and T-shirts.

In fact, the Easter Egg was the Doctor's way of communicating across time with Sally Sparrow. Together with her partner Larry Nightingale, she compiled a dossier on the strange events that centred on the deserted house Wester Drumlins. This letter was pushed through the door of their shop, Sparrow & Nightingale - Antiquarian Books and Rare DVDs, the night after she gave the dossier to the Doctor.

DON'T BLINK

DON'T BLINK

Oh, that was rude, wasn't it? And you were being so helpful. Saving my life, really. And Martha's.

But, like you guessed, we'd never met before. Well, I'd never met you. Well, not yet. And like I said, there was a sort of a thing happening. Four things in fact. And a lizard. And we only had twenty minutes until Red Hatching. Which is why the paper's got a bit yucky — sorry about that.

But I've read all the stuff you gave us now, so I know what's going on. Or rather, what will be going on. Or for you, what has been going on. And for Martha, what will have been going on. Probably. So — thank you, Sally Sparrow. Thank you so much.

And I'm sorry about your friend Kathy. And about poor Billy Shipton — though it looks like I'll get the chance to tell him that myself. But they both lived (or will have lived, have been living) full and happy lives.

When I said (will say) that time is a big ball of wibbly-wobbly timey-wimey stuff, I was (or will be) simplifying it rather. It's much more complicated than that.

And a heck of a lot more fun.

See you earlier

The Doctor

DON'T BLINK
DON'T BLINK
DON'T BLINK
DON'T

Missing for several days when the Lancaster bomber he was piloting failed to return from a raid, Reginald Arwell was presumed dead. His wife, Madge, was informed of her husband's loss just days before Christmas.

But, in one of the many unexplained incidents of the Second World War, Arwell's bomber did in fact return. Arwell landed it on Christmas Day, he and his crew unable to account for the time they had been gone. Even more bizarrely, the plane landed in the grounds of a house owned by Arwell's brother, where Arwell's own wife Madge and his children Lily and Cyril were staying at the time.

Also at the house was a new caretaker, standing in for Mr Cardew, who usually looked after the house in the absence of its owner, Digby Arwell. The absence of Mr Cardew is explained by this letter.

The TARDIS

Somewhere nearby.

Didn't check the road name. Sorry.

21st December (I think - I lose track rather)

Dear Mr Cardew

I am writing to inform you that your caretaking talents will not be required over the coming festive season. So you and Mrs Cardew can take a bit of time off and enjoy yourselves. In fact, why not go away on a little holiday? I've enclosed train tickets to Bournemouth (sorry, best I could do) and there's a map showing how you can get from the station to the hotel where there's a room booked for you. All paid for. Have fun.

Don't worry about the house, I'll look after it. I know Mr Arwell's sister-in-law Madge and her family are coming down for Christmas. I'll look after them too. Got a few ideas already, actually. I owe them a favour. At least, I owe Madge a favour. Long story, which involves lots of explosions, falling, and a back-to-front helmet. Let's save it for another day, shall we?

And when you get back, you won't know the place. Actually, that's a daft thing to say because you will know the place. But I have some ideas for a few improvements. I won't spoil all the surprises, but what about this: in the kitchen, taps for hot and cold water (boring) and - get this - lemonade! I know.

So you go off and have a jolly good Christmas with the Mrs. I've ordered you both the roast turkey for lunch because, well, because it's Christmas. And there's a present waiting for you at the hotel too. On no account open it.

Happy hols

The New Caretaker

(Also known as 'The Doctor'. And 'Get Off This Planet'. Probably.)

Nyssa of Traken first met the Doctor when he visited Traken towards the end of his fourth incarnation. After the death of her father - whose body was stolen by the Master - and the destruction of her home planet by an entropy cloud, Nyssa travelled with the Doctor and the TARDIS became her home.

In his fifth incarnation, the Doctor became particularly fond of the young orphan. Eventually, she decided to leave the TARDIS and work on Terminus, setting up a facility to cure victims of the horrific Lazar's Disease. This message was left for Nyssa on the Terminus main communications hub.

///THIS SYSTEM IS TO BE USED FOR TERMINUS INCORPORATED BUSINESS PURPOSES ONLY///

My Dearest Nyssa

You were so much younger when we first met, and I was a lot older. Funny thing, time. But during our travels together, we've been through so much - you and me, and Tegan. And Adric, too. And now Turlough, though I'm still not quite sure what to make of him.

But it won't be the same without you. You told me you'd enjoyed your time on the TARDIS, and yes we've had some fun. But there have been tragedies too, like poor Adric. And I have never forgotten that you joined us because of the death of your father, Tremas. He was a good friend, a good man. If I have managed to fill just some of the terrible hole in your heart left by his passing then I am thankful.

I cannot argue with your decision to stay on Terminus, much as I'd like to. You're right, the Vanir need your help if they are to break free of Terminus Incorporated and set up a proper hospital facility where they can treat and cure the Lazars. Talk to the Garm, he'll help now he is free of his programming. And Valgard and the other Vanir are good people really.

But you know that already. And I'm really telling you just to postpone the moment when I have to write this:

Goodbye, Nyssa. You are so brave and I wish you the very best of luck. I'm afraid you'll need it. But I know you will rise to the challenges ahead. You'll do good things, because you are a good person. There's a lot of your father in you.

Of course, Tegan will miss you dreadfully. And so will I.

The Doctor

Neither of these two notes ever actually existed. Preserved in the TARDIS's Telepathic Circuits, they were left by the Doctor for himself in each of the two worlds created by the so-called Dream Lord. The Doctor hoped his thoughts and ruminations would help him work out which world was real. They didn't.

In fact, both worlds were dreams and - like these notes - never existed.

Right Doctor, if you get back to this world then here's what I was thinking before I fell asleep last time. Or woke up. Whichever it is.

I really don't think this is Leadworth. Or even Upper Leadworth. I just don't.

Though I could be wrong.

And what about those Eknodines? I mean, are they even real? Or did I imagine them? Eyes in mouths, bit bizarre, but could be a real thing I suppose. The Vordervilt Tribe have tongues in their noses after all.

But no, not convinced. This has to be the dream world. Absolutely.

Unless it's the other one.

Hmmm.

I'll leave that with you, then.

The Doctor

OK, if you're back again Doctor, waking or sleeping, then here's the thing:

A star that burns cold. Is that even possible?

Because if not, then this is a dream.

Except, I know this TARDIS inside out and back to front and upside-down. So it's got to be real. Hasn't it?

Cold star... No - nopey-no-no-no. This is definitely the dream world.

I think.

Perhaps.

Though maybe not.

And now I'm talking to myself. On paper. Madness beckons.

The most holy and cherished relic of the Monastery of Det-Sen, in the foothills of the Himalayas in Tibet, is its Holy Ghanta. A small bell, the Ghanta is embossed with the symbol of a dragon.

It is said that if ever the Monastery is threatened, the Ghanta should be taken immediately to a place of safety. So when bandits attacked Det-Sen in 1630, the Master, Padmasambhava, entrusted the Ghanta to a stranger to smuggle away until the threat had passed. The stranger, who called himself only the Doctor, promised to return the Ghanta safely. But according to legend, it was almost three hundred years before the mysterious Doctor returned and fulfilled his promise. Legend also says that Pasmasambhava himself survived and waited for the Doctor...

When the Doctor did return, the monastery was again under threat - this time from the Yeti, which had apparently turned from gentle passive creatures into aggressive and savage beasts. With the return of the Ghanta to Det-Sen, the Yeti were defeated.

This letter is held in the Monastery's Library.

Padmasambhava

I am honoured and flattered to have been entrusted with your Holy Ghanta. I promise that I shall keep it safe, and do my best to return it to the Det-Sen Monastery once the current danger has passed.

I am sure that, with faith and determination, you and your fellow monks will resist the bandits who are attacking you. In other circumstances, I would of course stay and help. But I appreciate the trust you have put in me and I understand how important the Ghanta is to you.

And since trouble seems to follow me around, perhaps when I return the Ghanta to you, I shall have the opportunity to help you once again, and in a more direct manner.

I remain, sir, your friend and servant

The Doctor

143

This unsigned card was received by Jacqueline Tyler following the death of her husband Peter in a road accident. It was addressed to both 'Jackie' and Rose – the couple's baby daughter.

Deepest Sympathy

Dear Jackie and Rose

I am so sorry for your loss. Pete was a good man. Better perhaps than you will ever know. He was selfless and loving, and he put you - all of you - before himself.

But everything, and everyone, has its time. I know that more than most.

The memories won't always be this sad. Don't shy away from them. Cherish them. Remember and celebrate what you had, rather than mourning what you have lost.

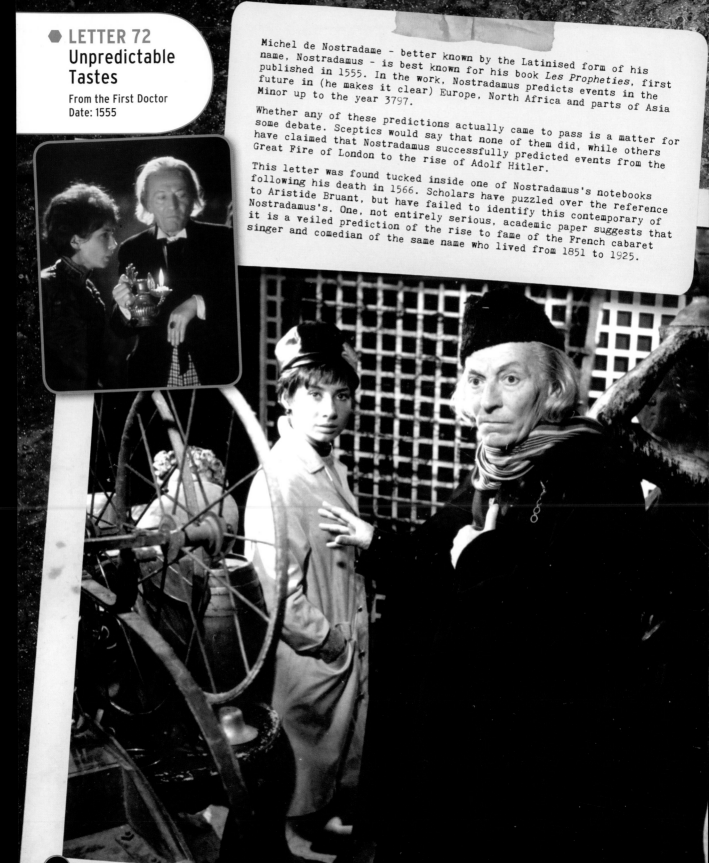

Michel de Nostradame – better known by the Latinised form of his name, Nostradamus – is best known for his book *Les Propheties*, first published in 1555. In the work, Nostradamus predicts events in the future in (he makes it clear) Europe, North Africa and parts of Asia Minor up to the year 3797.

Whether any of these predictions actually came to pass is a matter for some debate. Sceptics would say that none of them did, while others have claimed that Nostradamus successfully predicted events from the Great Fire of London to the rise of Adolf Hitler.

This letter was found tucked inside one of Nostradamus's notebooks following his death in 1566. Scholars have puzzled over the reference to Aristide Bruant, but have failed to identify this contemporary of Nostradamus's. One, not entirely serious, academic paper suggests that it is a veiled prediction of the rise to fame of the French cabaret singer and comedian of the same name who lived from 1851 to 1925.

Dear Michel

I am writing to thank you for your hospitality and most pleasant company, which helped to make the visit my granddaughter Susan and I made to Salon-de-Provence a most agreeable experience.

I was, as you will have realised, most intrigued by your 'Propheties', and I hope you will take into account some of the events which I divulged to you when composing further quatrains. As we discussed, I believe that when setting down such predictions, it is best to be vague, ambivalent, and generally as opaque as possible.

I would of course be grateful if you would keep to yourself my contribution to your work. I fear it might be rather frowned on in certain quarters. I shall say no more.

Please also convey our gratitude to your good wife Anne, who made us most welcome. Susan was especially taken with the handkerchief she embroidered for her. I have to say that I am not sure the long scarf she knitted for me is entirely to my taste, and cannot in all honesty see myself ever wearing it. Although, as we both know, it is foolhardy to make such predictions. That said, I shall of course treasure it. I have a wardrobe where I keep such items, including the recent addition of a rather battered hat that I was given by Aristide Bruant. Something else I do not think I shall ever have reason to wear.

But, as the saying goes, it is the thought that counts.

With gratitude again for your friendship and hospitality,

The Doctor

The Bank of Karabraxos was, until its destruction in a freak solar storm, the most secure financial facility in the galaxy.

Before the bank was destroyed, the Head of Security, Ms Delphox, uploaded all digital information held by the bank to secure off-world storage. That data included this application for a deposit facility, made just weeks before the solar storm struck.

To whom it may concern

In accordance with your rules and guidelines, I have completed an Application E-1-995W form which is attached to this virtual-mail message. According to Note 17b, this form must be accompanied by a short personal statement from the applicant detailing why they wish to open an account with the Bank of Karabraxos. So here it is:

I have recently come into a really quite ludicrous amount of money from the sale of rights in an idea for a 'sonic screwdriver' to the Magpie Electrical Company. Obviously some of this I have invested wisely, availing myself of the esteemed services of the Sil Financial Services Corporation of Thoros-Beta. But a considerable amount remains, which I should like to deposit – partly in cash and partly in the form of precious stones (including but not limited to diamonds, Andromedan bloodstones, madranite-one-five, Oolian, and PJX-one-eight).

Obviously, the Bank of Karabraxos is the most suitable establishment for me to make such a deposit, since it is the most secure bank in the galaxy and – according to your marketing literature – completely theft-proof.

On the assumption that my application will be successful, I also attach (as requested in Note 233d) a list of dates when I can be available for DNA sampling in accordance with your strict protocols. I apologise if some of the dates are in the past, as I get a little confused about these things. But I can still make those dates if that is convenient to you. But I suspect it isn't.

Yours Faithfully

John Smith (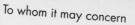 Architect)

During its long and largely profitable existence, the Interplanetary Mining Corporation (IMC) received numerous complaints about its heavy-handed approach to business. A high proportion of these came from colony worlds where IMC had allegedly coerced the colonists into moving on, before mining the planets for rich minerals and metals such as duralinium.

While IMC's official process called for such complaints to be registered through virtual-net comms traffic using a standard form, many colony planets and frontier worlds did not have access to such sophisticated facilities. This handwritten letter, received on 15 May 2472, is sadly not untypical.

To: Customer Services
Interplanetary Mining Corporation
PO Box 143-889-0
Earth Central

I am writing to complain in the strongest possible terms about the conduct of the personnel of IMC Survey Ship Four-Three under the command of Captain Dent.

I recently had occasion to observe Dent's methods first hand on the planet Uxarieus. As you will be aware (or if you aren't, you can easily check) this planet has been designated for colonisation. However, Dent and his crew feigned surprise at this, claiming that the planet was in fact also designated for mining due to a computer error.

Such an error might be plausible, except as you well know it never occurred. What is completely unacceptable is Dent's attempts to intimidate the colonists already on the planet. As well as the usual bully-boy tactics so beloved of large corporations like yourselves, Dent faked non-existent monsters using a standard mining robot equipped with fake claws, and even killed several colonists.

In fact, the situation deteriorated to the point where an Adjudicator from the Bureau of Interplanetary Affairs had to be sent for. Unfortunately, as the real Adjudicator, one Martin Jurgens, was replaced en route by an imposter who was only interested in an ancient Doomsday Weapon developed by the ancestors of the indigenous inhabitants of Uxarieus, his ruling cannot really be seen as fair and legal.

That aside, I demand that IMC conduct a full inquiry into Dent's conduct, with a view to reporting his criminal actions to the relevant authorities and offering suitable compensation to the colonists.

The Doctor

COMPLAINT REJECTED
REASONS: Completely implausible. Obvious time-waster.

Adipose Industries purported to offer a revolutionary new weight-loss product. But in fact, the excess fat from a customer's body was turned into baby Adipose creatures. Miss Foster, head of the company, was actually the alien Matron Cofelia, working for the Adiposian First Family to create millions of children. But once discovered, Matron Cofelia took drastic measures - intending the victims not just to lose weight, but to have their whole bodies converted into Adipose.

Following the Doctor's intervention and the death of Matron Cofelia, this communication was delivered to the Adiposian First Family by the Shadow Proclamation.

To the Adiposian First Family

I know what you were planning on Earth. I know what you did, even though you disposed of Matron Cofelia to try to cover your tracks. Sacrificing millions of humans just to create new children for yourselves out of their fat and bodily organs is unacceptable, even if that was just a contingency plan. No contingency can justify such action. I've sent word to the Shadow Proclamation, but I know as well as you do that without more evidence they'll do nothing.

But let me tell you – you don't get a second chance. Try anything like that again, and I'll stop you.

Just be thankful that your plan only partly succeeded. Be thankful for the new children you have. It's not their fault. They had no say in how they were born or at what cost. People died, and you had better make sure you remember that. Remember it every second of every hour of every day as you bring up those children.

Bring them up to be better than you.

The Doctor

One of the foremost experts in her field, Professor Emilia Rumford was surveying the Nine Travellers, a stone circle on Boscombe Moor, when she met the Doctor in the autumn of 1978.

Following her death, this letter was found tucked inside Professor Rumford's own copy of her 1976 book *Bronze Age Burials in Gloucestershire* – widely regarded as the definitive work on the subject. The letter is undated.

The book for which Rumford is best remembered is *Mud and Bones: The Accrington Priors Excavations*, first published in 1986.

Emilia

Thank you so much for all your help. Romana and K-9 and I couldn't have managed without you, you know. Really, we couldn't. Well, possibly. But it wouldn't have been half as much fun.

I'm so sorry to have messed up your stone circle. But I'm sure you'll enjoy surveying the Nine Travellers again, even though there still aren't nine stones. You could probably invent a good reason for that, or even tell the truth. But I suspect stories of blood-sucking alien creatures made of stone might not exactly enhance your academic reputation.

Speaking of academic reputations, I hope you don't mind but I had a quick look through your book 'Bronze Age Burials in Gloucestershire'. It really is quite detailed, isn't it? Though I'm afraid some of the details aren't entirely true. I couldn't help spotting the reference to the Peruvian temple designs in Chapter 7, for example. Well, they were influenced by the Exxilons, which rather invalidates the subsequent argument. And as for the second paragraph on page 197 - no. Absolutely not. And I was there, so you can take my word for it.

But generally it's a good effort. Well done. I'm hoping to find a copy of your book on the archaeological significance of the Accrington Priors excavations. But I don't think you've written it yet.

But you will. And it will be marvellous.

The Doctor

THE LAST C

The town of Tombstone in Arizona is best known for the gunfight at the OK Corral that took place in 1881. Prior to that, one of the main participants, John Henry 'Doc' Holliday, a graduate of the Pennsylvania College of Dental Surgery, ran a dental practice in the town.

Perhaps because it was from his first customer in the new business, Doc Holliday kept this letter, despite its rather scathing description of the treatment the patient received.

Sir

While my toothache does indeed seem to have been mitigated by your treatment, if one can call it that, I feel I must protest at your level of what I believe is usually referred to as 'customer service'. Especially considering that I was, I believe, your first customer, I should have expected a rather greater degree of comfort and hygiene. I do not believe that reaching into the patient's mouth with a large pair of pliers which seem to have been used to straighten spurs and remove nails from horseshoes is perhaps the most sanitary of methods for removing a troublesome tooth.

While I cannot reasonably fault you for the lack of anaesthetic, given that it is only 1881, I am not sure that the offer of a 'rap on the cranium with this six-shooter' is quite the level of care one might expect. Nor was the offer of a 'slug of rattlesnake oil', whatever that may be. Though I did notice you were careful to imbibe a quantity of the substance yourself before commencing the operation.

In short, sir, while I commend you on the outcome of your service, I do think you could usefully pay more attention to the experience you offer as a whole. I doubt very much that I shall return for further treatment.

Still, I suppose I can't really complain, as you were kind enough not to charge me. I hope Miss Kate is pleased with the gift you promised to make her of my tooth.

Good day to you, sir.

The Doctor

Approximately two billion light years from Earth, Apalapucia is a beautiful planet of soaring spires, silver colonnades, and the famous mirrored Glasmir Mountains. But the inhabitants fell victim to an outbreak of Chen-7, which rapidly kills races with two hearts like the Apalapucians (and indeed the Time Lords).

The Two Streams Facility was established using temporal technology to extend each victim's last day of life into an apparent lifetime. Trapped inside the facility, the Doctor's friend Amy Pond waited for the Doctor and her fiancé Rory to rescue her. But while a few hours passed for them as they tried to get to her, 36 years passed for Amy. Her notes were later discovered by Handbot robotic nurses patrolling the facility.

Rory, Doctor

I can wait.

I've waited before. I'll probably wait again. I'm good at waiting — the girl who waited, that's me.

But really, I didn't think you'd keep me waiting **this** long.

At first, I worried that you might be a while. Then I worried something had happened to you. Now I worry that you may not come at all. Have you any idea how long —

No. Deep breath. Hold it together, Pond. You're on your own now. Just Amy and Interface and Rory.

Not you, Rory — the other Rory. I've waited so long I made a new one. He's not quite like you. Quieter for one thing. That's good and bad. He doesn't answer back, but he doesn't tell me he loves me. I miss your voice. I miss your everything.

Do you miss me?

If so — where the hell are you?

You told me to wait, Doctor. And not for the first time either. But it's not just a few years this time, is it, Raggedy Man? Not just till I'm grown up.

It's till I'm grown old.

Maybe it's until I die.

In 1885, Herbert George Wells was holidaying in Scotland, not far from Inverness, when he had a life-changing experience. The young man had always been interested in writing and in 'speculative fiction', but his brief travels in an actual time machine provided the raw material for the novels he would later write. *The Time Machine*, his first novel, was published in 1895.

This letter was waiting for H.G. Wells on his return from Scotland.

Who was it who said 'A writer is someone who can't not write'?

No, I don't expect you to know, Herbert. After your time, I'm sure.

And there's the thing – Time. You're going to write about it, I know you are. Because you've just had the most incredible experience (with the Timelash and the Borad and everything) and met the most amazing people (me, chiefly), and you can't not write about it. It's what you do.

It's who you are.

So just be careful. That's all.

Don't tell it exactly how it was. If you do, I could get into all sorts of trouble, to say nothing of potential damage to the time lines. So dress it up a bit. Improvise. Embellish. If it's possible, make the story even more exciting and thrilling.

Maybe, and this is only a suggestion, split it up a bit. We've done enough in the last day or two to fill several novels. Time travel in one, experimenting with the nature of life in another. And that's just for starters.

It might take you a while, but you'll do our exploits justice. I know you will.

Peri says 'Hi' (make allowances – she's American)

The Doctor

UNIT

FORMERLY THE UNITED NATIONS INTELLIGENCE TASKFORCE AND NOW THE UNIFIED INTELLIGENCE TASKFORCE, UNIT IS A SECRET PARAMILITARY ORGANISATION WHOSE MISSION IS TO DEAL WITH THE ODD AND THE UNEXPLAINED - ANYTHING ON EARTH OR FROM BEYOND...

The Doctor first encountered UNIT soon after its formation. He already knew Brigadier Lethbridge-Stewart, the officer commanding the UK contingent of UNIT. They first met battling against the Great Intelligence and its robot Yeti in the London Underground when Lethbridge-Stewart was a colonel. The first full UNIT operation was against another alien threat the Doctor had encountered before - the Cybermen.

When he was exiled to Earth by the Time Lords, the Doctor's arrival coincided (perhaps by design) with an attempted invasion by the Nestenes who planned to use killer Autons to take over the planet. Again, the Doctor worked with UNIT to defeat the threat, and subsequently joined the organisation as its Scientific Adviser.

This appointment gave the Doctor a base for his time on Earth. More than that, it provided him with a home. The members of UNIT with whom the Doctor had most contact became firm friends, particularly the Brigadier, Captain Yates, Sergeant Benton and the Doctor's assistants Dr Elizabeth Shaw and Jo Grant.

As well as helping to combat alien menaces like Daleks and Cybermen, Axons and Dæmons, the Doctor also found himself up against an old friend - the renegade Time Lord known as the Master. Warned of his arrival on Earth by the Time Lords, the Doctor and UNIT soon found the Master to be their number one enemy. Even his arrest and subsequent isolated imprisonment did not end the problem, as the Master soon escaped, allying himself with the prehistoric Sea Devils and helping them in their attempt to 'reclaim' their planet.

With his exile ended, the Doctor slowly distanced himself from UNIT - more so after he regenerated into his fourth incarnation. But he has maintained ties with the organisation, and come to their aid - willingly or through necessity - on various occasions. He defeated the Sycorax, was recalled by Martha Jones to battle Sontarans, and helped to fight off deadly flying stingrays. He dealt with the so-called Slow Invasion, and again encountered the Master - albeit in a very different form - together with the Cybermen when investigating the mysterious 3W organisation...

Bureaucracy being what it is, UNIT regulations stated that the Brigadier needed to be able to produce a valid application form for all UNIT personnel serving under his command. This also applied to the Doctor, who was somewhat less than impressed at being asked to fill in an application form after he had already begun work as UNIT's Scientific Adviser.

Nevertheless, UNIT still has the completed form on file.

CIVILIAN TITLE (MR, MRS, DR, ETC) OR RANK IF CURRENTLY SERVING: Doctor - obviously

FULL NAME: The Doctor

NATIONAL INSURANCE NUMBER: I have no idea what you're talking about.

ADDRESS: The TARDIS, care of Brigadier Lethbridge-Stewart, UNIT HQ

CONTACT TELEPHONE NUMBER: Ask the Brigadier.

NATIONALITY: Gallifreyan

CURRENT OCCUPATION: Oh for heaven's sake - this is ridiculous.

HAVE YOU BEEN ISSUED WITH A TEMPORARY UNIT PASS (YES/NO)? Yes.

IF SO, STATE THE PASS NUMBER: No idea - I never carry the thing.

POST FOR WHICH YOU ARE APPLYING: I'm not. I'm already doing it. Do we really have to go through this bureaucratic nonsense?

ACADEMIC QUALIFICATIONS:
You wouldn't believe me if I told you.

STATE BRIEFLY WHY YOU BELIEVE YOU ARE SUITABLE FOR THE JOB:
Because the Brigadier asked me. That should be good enough for you pen-pushers. But if not, I can assure you I am uniquely qualified to deal with anything UNIT is asked to handle.

GIVE BRIEF DETAILS OF ANY RELEVANT EXPERIENCE YOU HAVE:
Helped Lethbridge-Stewart defeat the Great Intelligence and its robot Yeti. Stopped the Cybermen from invading Earth. Nestene Autons - remember them? Daleks, Ice Warriors, Quarks, the Terrible Zodin. This is a complete waste of time.

REFERENCES – PLEASE CAN YOU PROVIDE THE NAMES AND CONTACT DETAILS FOR TWO PEOPLE WE CAN ASK FOR REFERENCES. IDEALLY, ONE SHOULD BE A PREVIOUS EMPLOYER AND ONE A CHARACTER REFERENCE.
No I can't.
Talk to the Brigadier. That's Brigadier Lethbridge-Stewart of UNIT. He can sort this out.

U.N.

Although the Doctor became increasingly absent from UNIT, his laboratory and personal effects were not cleared until UNIT HQ was moved to its current location. Amongst the papers that were retained is this letter of resignation.

It is unsigned, which suggests that the Doctor never completed writing it. Certainly it was never sent to Brigadier Lethbridge-Stewart.

Lethbridge-Stewart

I am writing to express my outrage at your recent actions. I am sure you were only carrying out your orders, the usually military excuse. But nonetheless, I feel I can no longer be part of an organisation that has been complicit in the destruction of an entire race of intelligent beings. I accept that the Silurians did try to use a plague to wipe out the human race first, but I am sure that in time and with patience a negotiated settlement could have been reached.

I therefore offer my resignation from UNIT.

Not that I was ever really an employee.

But I admit it will be hard to find another organisation that can provide the latest modern scientific equipment for me to continue my work on the TARDIS. Or the generous level of funding. Or a competent and qualified assistant. Or will turn a blind eye to the fact that I don't officially exist on this planet.

It's a shame because you know I was actually beginning to enjoy myself.

Project Inferno, as it was nicknamed, was designed to drill through the outer layer of the Earth's crust. The plan was to tap into a powerful energy source called Stahlman's Gas, named after the instigator and lead scientist of the project, Professor Stahlman.

As a government project, security was handled by UNIT. Also on site were the Doctor and his assistant Elizabeth Shaw. As well as monitoring the progress of the drilling, the Doctor took the opportunity to use the project's nuclear reactor to provide power for his attempts to get his TARDIS working again.

Stahlman, a difficult man to work with at the best of times, objected to the Doctor's presence and to his demands for access to the reactor - as can be seen from this note to Stahlman from the Doctor. Stahlman himself was killed in the series of catastrophic accidents that led to Project Inferno being abandoned.

Professor Stahlman

I understand that you have once again disconnected the power from the main reactor to the hut where I am working. You may be the nominal head of this project, sir, but as you know I am engaged on vital scientific work for UNIT.

Since making demands of you seems only to aggravate your liver condition, I would request in the strongest terms that you restore power to my experiments at once. My work draws only a fraction of the main reactor's output which obviously won't interfere with the drilling. In the absence of a reply, which, judging from your previous lack of response seems highly likely, I shall assume your agreement and send my associate Miss Shaw to reconnect the systems.

The Doctor

This brief summary of the threat posed by the renegade Time Lord the Master is still held on file at UNIT. It is undated, but is believed to have been written shortly after the Daffodil Incident, in which the Master allied himself to the Nestenes in an attempt to invade Earth. Their plan was to kill thousands of people using deadly plastic daffodils and create a panic during which the Nestenes could invade. With the Doctor's help, the plan was foiled by UNIT.

Brigadier

You asked me to let you know just how great a threat the Master really is. The simple answer is that he poses the very greatest threat – not just to this planet, but to the whole universe.

I've already told you that he is a Time Lord, as I am. A renegade. But unlike me, he craves power, and he'll do anything to get it. He may be an unimaginative plodder, but he has a talent for destruction and he revels in carnage.

As well as being a master of disguise (no pun intended), he is also a formidable hypnotist. Anyone, or almost anyone, could fall under the Master's influence. Only people with an exceptionally strong will can resist.

What else can I tell you? He's arrogant and vain. He delights in the suffering of others – he'd delay an execution to pull the wings off a fly. But with it, he's absolutely charming when he wants to be.

He's also totally self-centred. He'll ally himself with anyone – or anything – to get what he wants. But he'll betray them on a whim or to save his own skin.

Oh and he has a working TARDIS. So he could pop up anywhere at any time.

I think that about covers it.

The Doctor

With Brigadier Lethbridge-Stewart busy with other matters, the Doctor took the unusual step of going directly to the Ministry of Science for support in his efforts to assess the potential threat posed by the Keller Process.

The Doctor never did get in touch with the Minister following the demonstration to which he refers in this letter. Events moved too quickly for that, involving not just the Keller Machine but also the Master, an international peace conference for which UNIT was providing security, and UNIT's clandestine disposal of a nuclear-powered Thunderbolt nerve gas missile.

Dear Minister

You asked me to put my concerns regarding the so-called 'Keller Process' in writing, so here they are.

You may recall that we spoke on the telephone last week, but to remind you I am the Scientific Adviser to UNIT, reporting directly to Brigadier Lethbridge-Stewart. As such I feel I am well qualified to assess the possible implications of the process.

As you know, the Keller Process uses a machine – the Keller Machine – to extract from the brains of criminals and social misfits the negative cerebral impulses that govern their antisocial behaviour. These impulses are stored, as I understand it, within the machine. It doesn't take a genius to work out that at some point the machine will be full of these negative impulses. I believe that a total of 112 cases have been processed so far in Keller's native Switzerland, so I have to conclude that the machine must be reaching capacity.

But in addition to the question of what to do with these negative impulses and how they can be safely stored or neutralised, there is another potential problem. Keller maintains that his process leaves the subject as a rational, well-balanced individual. However, I can find no evidence of this. In fact, there appears to be no record at all of the names of the people who have so far undergone the process. The Swiss authorities put this down to concerns over personal privacy. I suspect there may be more to it than that.

Which is why I intend to be present at the demonstration of the Keller Machine this week at Stangmoor Prison. I should like to remind you, Minister, that UNIT was set up to deal with new and unusual menaces to Mankind. In my opinion, the Keller Machine poses just such a threat. I shall be in touch again once I have seen the machine demonstrated and assessed the danger. I shall of course expect your full support for whatever action and precautions I believe are necessary.

The Doctor

Scientific Adviser to UNIT

LUNA UNIVERSITY

Professor Arthur Candy, Head of Archaeology and Psycho-History at Luna University, is noted among other things for a lecture entitled *Doctor Who?* Sadly, the text of the lecture no longer seems to exist, and there is much speculation as to what Candy might have said, and indeed what the subject of his lecture actually was. Candy himself was uncharacteristically reticent on the subject.

Although Candy kept meticulous research notes for all his other lectures, along with the final texts, there is no trace in his archives of that particular lecture. The only paper that could possibly have relevance is this letter.

River Song was indeed enrolled on Candy's course and graduated with distinction, eventually becoming a professor in Candy's department. Her long and varied - and controversial - career is well documented elsewhere.

Dear Arthur

You know I said that if you stopped telling everyone about me and going on about how I'm a 'Complex Space-Time Event' I'd leave you alone and you'd never hear from me again.

I lied.

I do that, I'm afraid. Bit of a habit. So here I am.

But it's an easy one this time. No one needs to get hurt or damaged or dumped in Birmingham in the mid 1990s for months on end. It's about one of your students. Or rather, it isn't.

Because she isn't a student yet. But she's applied. And she'll be amazing. A-maze-ing. She will. She just needs you to accept her on to your archaeology course at Luna University. As soon as you can.

Slight problem - she probably doesn't know she's applying yet. Actually, she's had a bit of a setback and is being cared for by the Sisters of the Infinite Schism. Nothing to worry about. Soon be back on her feet. But if you can just drop off an application form with them to pass on to her, together with a note along the lines of 'Here's the form you requested, no really you did' that'd be terrific. Thanks.

Oh, and her name is River Song. Well, it isn't really. It's Melody Pond. But River Song will do for admin purposes. And other purposes too.

So there you go. Simple request. You won't regret it. Promise.

And then I really will leave you alone. Probably.

The Doctor (Complex Space-Time Event)

12340 SE 345882   No.0013 658 2568

LUNA UNIVERSITY

RIVER SONG
Archaeology

Professor Arthur Candy

647568127969O-377588583524324S

Sergeant Michael Smith was attached to the Intrusion Countermeasures Group, under the command of Group Captain Gilmore during the so-called Shoreditch Incident in 1963. 'Countermeasures', as the group was known for short, was in many ways a precursor to the later UNIT organisation, combining both military and scientific expertise.

Smith was killed in the line of duty, although for reasons that were never divulged he was not accorded a military funeral. A private ceremony and burial took place at the local church. Afterwards, a Book of Condolence was presented to Smith's mother, who ran a small bed and breakfast in the area. The Book of Condolence contains the following entry.

Dear Mrs Smith

I'm sorry that my friend Ace and I couldn't stay for Mike's funeral. I'm afraid I don't like funerals – like burnt toast and bus stations. I don't mean to trivialise things – we were thinking of him, and of you.

Mike was a brave young man who did what he believed was for the best. He wasn't always right – well, who is? But in Mike's case, his mistakes got him killed.

I hope your memories of him will be happy ones. I'm sure he did a lot of good things in his all-too-short life. Don't think badly of him, any more than I do. Don't think badly of anyone. Tolerance is a great healer and understanding can right all manner of wrongs.

With the deepest sympathy

The Doctor

For many years, 'Professor' Herbert Clegg toured the regional theatres of Great Britain with a mind-reading act. Working with an assistant, he purported to guess cards, and discern personal information from members of the audience.

Although Clegg's act undoubted started out as just that - an act - experts are divided as to whether the man did actually develop some level of clairvoyance and extrasensory perception. This theory was lent credence by the accounts given by his assistant after Clegg's sudden death from a suspected heart attack. The assistant suggested that Clegg also had some psychokinetic ability, and could move small objects by the power of his mind.

The truth, of course, will never be known. Herbert Clegg collapsed and died while visiting UNIT Headquarters midway through what became his final tour. This note was in Clegg's coat pocket when he died.

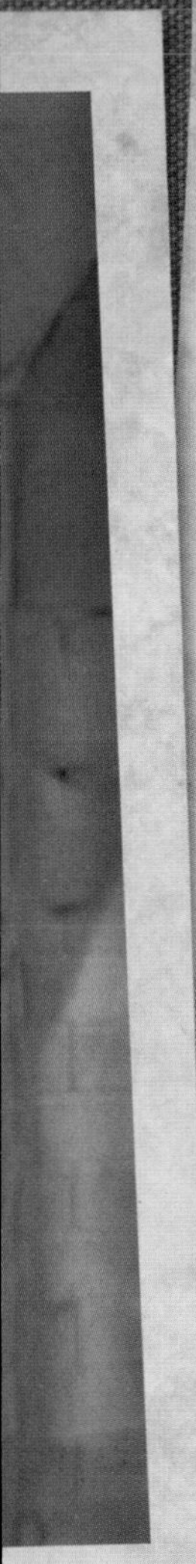

Professor Clegg

I was most taken with your act this evening, and wonder if you could spare a few moments tomorrow to meet with myself and my colleague Brigadier Lethbridge-Stewart?

I'm afraid the Brigadier was rather more taken with the physical attributes of the preceding act, Miss Scheherazade. But it was the more cerebral – indeed, sensory rather than sensual – nature of your own presentation that appealed to me. I have seen a great many so-called 'mind readers' over the years, but your own skills impressed me far more than most.

If I could persuade you to join us tomorrow morning, I should be most grateful. On the assumption that you accept my invitation, the Brigadier will arrange for a car to collect you from the stage door at 11am.

I look forward to speaking with you tomorrow.

The Doctor

The
Brooke House
Theatre

● LETTER 88
A Masterful Defence

From the Third Doctor
Date: Classified

Following his attempts to summon Azal, last of the Dæmons, in the village of Devil's End, the Master was captured by UNIT and put on trial in front of a specially convened court. Various UNIT personnel were key witnesses, including Brigadier Lethbridge-Stewart, Captain Yates, Sergeant Benton, Miss Josephine Grant, and the Doctor.

In addition to his statement in front of the court, and subsequent cross-questioning, the Doctor submitted a written statement which ran to 132 typewritten pages. The complete statement remains highly classified, and copies are held on file at both the Ministry of Justice and UNIT Command in Geneva. The introductory cover page to that statement is reproduced here.

In addition to my personal statement to the court, I submit this written testimony.

In the following pages, I shall enumerate some of the crimes of which the Master is undoubtedly guilty and give my own personal account of relevant events.

While it is clear that the Master has caused death and suffering on a vast scale, I should also like to take this opportunity to plead for clemency. He deserves the very harshest punishment, but I firmly believe that imprisonment, and the resulting loss of liberty, is a far more effective and humane approach than execution.

I am fully aware that several of the witnesses, and indeed members of the judiciary, have argued strongly for the death penalty to be reinstated for the Master. But I do not believe that any society that calls itself civilised can in all conscience condone such an act – even in this most extreme of cases.

Is it possible that the Master might reform, might see the error of his ways? Well, anything is possible, though I have to admit I feel it is highly improbable. But if we were to execute him for his crimes, then surely we are descending to the same level, rather than demonstrating the highest regard for justice.

Of course, there remains the problem of where such a criminal should – or indeed could – be incarcerated. I would suggest a special facility with the very highest levels of security, staffed by people who are demonstrably immune to the Master's formidable powers of persuasion and hypnotism. For the record, I am more than happy to advise on how such a facility might be established and run.

With that said, the following pages detail events on this planet in which the Master was involved. In order, I shall give my evidence on the (second) attempted Nestene invasion, the business of the Keller Machine and the Stangmoor Prison incident, the arrival of Axos and the offer of Axonite, and the recent events at Devil's End which led to the apprehension of the Master.

Immediately after regenerating into his fourth incarnation, the Doctor became involved with the so-called Giant Robot Incident. Experimental Prototype Robot K-1 had been developed and built by Professor J. P. Kettlewell, under the auspices of the Think Tank organisation. Think Tank had been set up by the UK government to research into the frontiers of science, but by this point was under the clandestine control of a fanatical extremist group called the Scientific Reform Society.

The SRS used Kettlewell's robot in a plan to blackmail the international powers into acceding to their demands for reforming world government. When the robot's actions conflicted with its programmed Prime Directive to serve humanity and never harm it, the Robot went to find its creator. Kettlewell in turn alerted the Doctor to the robot's presence at his house.

The Doctor agreed to meet Kettlewell and his robot. But suspecting (correctly as it turned out) that the professor might be working with the SRS, he left this typed note in the UNIT laboratory for his friend Sarah Jane Smith to find.

Sarah

Professor Kettlewell tells me that he has the robot hidden at his house.

Gone to meet him.

P.S. It is of course possible that this message is a trap. If it is I can deal with it.

P.P.S. I am leaving this note in case I can't.

At the time of the attempted Cyberman invasion organised with the collusion of Tobias Vaughn of International Electromatics, Professor Edward Travers was with his daughter Anne in the USA. While they were away, Travers lent his flat – where Anne had a laboratory – to Anne's former tutor, Professor Watkins.

Although Watkins himself was caught up in Vaughn's plans, the Doctor and UNIT enlisted his help and that of his niece Isobel Watkins in defeating the Cybermen.

This letter was left in the flat, and found by Travers on his return some 11 months later.

Travers, my dear friend, how are you?

Jamie and I came looking for you with Zoe (you've not met her yet), as I was hoping you could help me give the TARDIS a bit of an overhaul – it's long overdue. But we found you'd gone off to America for a year with Anne. How exciting.

Mind you, things were quite exciting here too. I don't know if you're aware, but your good friend Professor Watkins was caught up in that nasty business with International Electromatics, working for Tobias Vaughn, no less. Isobel, his niece, was very helpful and once we'd met Watkins we soon discovered that some old adversaries of mine, the Cybermen, were planning to invade this planet.

The Cybermen are robots, like the Yeti. Or, actually, not like the Yeti. And not really robots. I'll explain it all to you when we do finally meet up, which I hope won't be long.

But we ran into another old friend, too. You remember Colonel Lethbridge-Stewart who was in charge during those shenanigans with the Great Intelligence in the London Underground? Well, he's now in charge of something called UNIT, which is a multinational military force formed to deal with exactly this sort of thing. Mind you, they still have a lot to learn. Wouldn't have got very far without my help, I can tell you.

Anyway, without your assistance, I suppose I shall have to postpone my TARDIS overhaul. The poor old thing is getting on rather, but I'm sure she'll manage for a few more centuries yet.

Hopefully I'll catch up with you before that!

The Doctor

Sir Reginald Styles was instrumental in persuading world powers, and particularly the Chinese, to hold another peace conference after the events of the so-called Keller Affair resulted in the first conference in London being abandoned. The reconvened conference was to be held at Auderly House, a government-owned stately home approximately fifty miles north of London.

But hours before the conference was due to start, Auderly House was destroyed in a huge explosion. The official report into the incident attributes the blast to a leaking gas main. The truth is altogether more sinister.

Amongst the paperwork preserved from the historic – and successful – conference, is this letter delivered to Sir Reginald Styles by UNIT messenger just minutes before the conference finally began at its new location, a government facility just outside Hemel Hempstead.

Sir Reginald

You told me that you are your fellow delegates are confident the peace conference will be a success because you know what will happen if you fail. I'm sure you think that's true. But I've seen it happen, and I can assure you it is a lot worse than you can possibly imagine.

A hundred years of nothing but killing. Seven-eighths of the world's population wiped out, and the survivors reduced to living like animals in the ruins of your planet. And that's not the worst of it. Not by a long way.

With the human race weakened almost to the point of extinction, that is when your planet will be invaded. I can't tell you too much about the possible future. There are rules that even I can't break. But if the 'ghosts' who came to your house before you left for China seemed desperate, it is because they were. As desperate as anyone ever gets. They were wrong in their interpretation of events, but they would certainly have resorted to your murder to prevent the future from which they came ever happening.

Ultimately, those same people saved your life, and the lives of the other peace conference delegates. One of them, Shura, sacrificed his own life so that you could live. So that you could forge a peaceful future in which the human race survives and thrives.

Don't let his sacrifice be for nothing.

The Doctor

While the main buildings of Deffry Vale High School were destroyed in an explosion apparently caused by a faulty heating system, the administration block was largely undamaged. Amongst the papers retrieved from the school during the subsequent UNIT investigation was this letter addressed to the relatively new head teacher, Mr Hector Finch, who sadly died in the accident.

Strangely, there is no record of a teacher named Doctor John Smith registered with the Local Education Authority, neither does the 'Schools Supply Cover Service' seem to exist.

SCHOOLS SUPPLY TEACHER COVER SERVICE

Dear Mr Finch

Miss Morrison has asked that we arrange a cover teacher for her. I realise that a replacement at such short notice is unusual except in case of sudden illness, but the circumstances are exceptional. Although she does not as a matter of routine participate in the National Lottery, Miss Morrison has, due to circumstances that need not concern us here, found herself in possession of the winning jackpot ticket. As a result, she has decided to take a short leave of absence from Deffry Vale High School while she decides what to do with her new-found wealth. And also take a holiday. I'm sure she will contact you herself on her return.

In the meantime, as the approved Supply Agency for your region of the county, we are sure you will find Miss Morrison's temporary replacement more than satisfactory. John Smith is, if anything, over-qualified for the position of Physics Teacher, holding as he does a doctorate as well as several other degrees from various highly prestigious universities. You will find him enthusiastic, diligent, talented, and a joy to work with. I can tell you with all honesty that he is one of the very best teachers we have on our books.

I should also mention that we have allocated a new Kitchen Assistant to your school as the catering facilities seem to be slightly under-staffed at present. Miss Tyler is also highly qualified and will fulfil the post admirably.

If you have any questions please don't hesitate to contact us.

James Robert McCrimmon

(Head of Resource Allocation)

The internal UNIT memo reproduced here is typical of the communications sent by the Doctor during his time as UNIT's Scientific Adviser. This particular memo was sent to the head of UNIT's Scientific Supplies Section shortly after UNIT's first encounter with the Master.

It would appear from the memo that equipment the Doctor deemed vital to analysing the threat posed by the Master and his Nestene allies was not immediately available as it was not held in stock. In other, similar, memos the Doctor complains to various officials that the equipment he needs has not been invented yet.

SUBJECT: STOCK HELD IN UNIT STORES

Mr Campbell

I understand from my assistant Miss Grant that you are in charge of UNIT's Scientific Supplies Section. While she tells me that you personally have been most helpful in expediting the delivery of the equipment I ordered recently, I have to say that it does seem the UNIT stores are woefully ill-equipped. I find it hard to believe that not only could you not immediately supply me with the Electrode Unit I asked for, but that you did not even have a Scanning Molecular Structure Analyser in stock and had to send out for one.

I have raised this matter with Brigadier Lethbridge-Stewart, and he agrees that the situation is simply unacceptable. I cannot be expected to work without proper equipment.

Therefore, I shall shortly be sending you a list of what I consider to be essential items to be kept in stock at all times. It will include everything from a Diothermic Energy Exchanger to a Lateral Molecular Rectifier. Rather primitive, I know, but it's probably the most advanced equipment there is at the moment. If you have any budgetary issues once you've seen the list, kindly raise them with the Brigadier.

Any other issues, please contact myself or Miss Grant direct.

The Doctor

Josephine Grant worked as the Doctor's assistant for most of his time as UNIT's Scientific Adviser. Following the Global Chemicals incident in South Wales, Miss Grant resigned from UNIT to marry Dr Clifford Jones, a Nobel-prizewinning environmental scientist devoted to finding alternative energy and food sources.

Their engagement party was held at the Wholeweal Community (locally nicknamed the Nuthutch) run by Jones in Llanfairfach. It was a double celebration as, with the help of Jones and his colleagues, UNIT had just defeated the plans of the megalomaniac supercomputer BOSS which ran Global Chemicals, as well as dealing with an infestation of giant maggots mutated by dumped chemical waste.

The Doctor slipped away from the party early, returning to UNIT HQ in London. As well as the gift of a rare blue crystal from the planet Metebelis Three, the Doctor also sent this note to Jo Grant the following day.

My Dear Jo

Forgive me for slipping away. I'm not really one for goodbyes, as you know.

I suppose also that I'm sad our partnership is coming to an end. We may have
started off inauspiciously, but you have become invaluable to me, you know.
Another few years and I might even have made a scientist of you.

But every ending is a new beginning. And if I have to lose you to someone else,
then you could not have chosen better than Cliff Jones. He'll look after you, and
I know you will look after him too. And who knows, maybe your new doctor
really will turn you into a scientist.

I'm sure your trip up the Amazon will be a big adventure. Perhaps as big as some
of the adventures you and I have shared. But the biggest adventure of all will
be your new life together – as thrilling and as exciting as anything you've done
before.

I shall miss you. But I take comfort in the knowledge that despite all our happy
times together, for you the best times are still to come.

The Doctor

Found among the personal papers of Queen Victoria after her death, this letter believed to have originated in 1879, has baffled experts. While it makes no explicitly anachronistic references, the turn of phrase is most certainly more modern than can be explained.

The Torchwood Estate did indeed exist in 1879. The Scottish residence of Sir Robert MacLeish, it gave its name to the Torchwood Institute, founded by Queen Victoria and regarded (by those who are privy to its existence) as a forerunner to UNIT. However, no Powell Estate as referred to in the postscript has ever been identified in the area.

THE TORCHWOOD ESTATE

Your Majesty

Well, banishment — that seems rather harsh. I mean, all right, so you were attacked by a werewolf and a load of mad mistletoe-wielding monks, but still... It's not like it was our fault. I'm sorry, but while the knighthood is all well and good and gratefully received, exile seems a bit steep.

Ironic too, as I was once exiled to your kingdom. Mind you, in strictly chronological terms that hasn't actually happened yet. But even so.

Still, you're upset and confused and scared and definitely not amused — that's understandable, even for a Queen-Empress. I meant what I said, though. Prince Albert, your husband, he had that diamond cut to save your life. It's obvious that you loved him — still love him — very much. It's equally obvious that he loved you very much too. He'd have given his life for you, if he still had it to give.

I doubt you'd have banished him for it.

And actually, this isn't the first time I've saved your life, or your empire. But don't worry about it — banished and exiled or not, I'd do it again. I probably will.

Maybe we'll meet again, I don't know. In fact, if we do, you probably won't know anyway. People don't always recognise me when we meet again.

Despite everything, I remain your humble subject (well, sort of)

Sir Doctor of TARDIS

PS — Dame Rose of the Powell Estate says hello. She's not too chuffed about the banishment thing either, but she'll get over it. Well, probably. Well, maybe. Well...

The residents of the small village of Devil's End rarely speak about the events that led to the destruction of their church and the apprehension of the Master. Although millions watched Professor Horner's notorious, and tragic, opening of the ancient barrow known as the Devil's Hump when it was broadcast live on BBC3 as part of the *Passing Parade* series, what actually took place remains shrouded in mystery.

The church was eventually rebuilt, helped in no small part by a substantial contribution from a Doctor John Smith.

The Reverend Mr Magister

Dear Mr Carmichael

I am writing to you in your capacity as the Chairman of the Parochial Church Council for the village of Devil's End. Normally, I would have written to the incumbent vicar, but I know that Canon Smallwood was forced to retire through ill health and his successor, the Reverend Mr Magister, has since moved on under something of a cloud...

I recently visited Devil's End, staying at the Cloven Hoof, which is a most accommodating establishment. We may even have met – my friend Miss Grant and I are associated with UNIT and arrived the evening that Professor Horner opened the barrow. Perhaps you were one of the Morris Dancers or joined in the dance round the Maypole. Forgive me if I don't recall everyone's names.

But to come to the point, knowing that the village church was badly damaged in the events of those few days, and being not only aware of the circumstances but somewhat involved myself, I do feel a little responsible. Therefore, I should like to make a small contribution to the fund, and I hope you will accept the enclosed cheque as a donation towards the costs of repairs. I think it's valid, although I have to confess I use money so rarely I'm rather out of touch with how the banking system works in the late twentieth century.

I hope I shall find time to return to Devil's End some time soon to see how the renovation work is progressing.

With all best wishes

Doctor John Smith

DEVILS END

This letter was found lodged inside a chimney during renovations to a house in Baker Street, London. From its condition, it would seem to have been there for over a century, and yet the paper is of a modern composition.

Tests on the paper to determine how it could have survived unburned in a working chimney were carried out by Professor Malcolm Taylor of UNIT. They revealed that it has been treated with a fire-retardant coating which has so far defied analysis.

Thank you

Dear Santa

Well, there's something I didn't think I'd ever write. Not ever. And for me 'ever' is a very long time indeed.

I told Clara I don't know who to thank. That was a lie. Rule one – or it always used to be – the Doctor lies. Big time.

But, and this is definitely not a lie, I owe you a debt of gratitude. It's not often I have to admit that I needed someone else's help. But I did, and it was you who were there.

Have I been naughty or nice? You know, I really have no idea. Either way – Best Christmas Present Ever. Thank you.

But seriously – you should give up on the tangerines. No one likes them. No one at all. Signature gift or not, it's time to move on, Big Fellah.

And a happy Christmas
to all of you at home

The Doctor

PS – Do I really stick this
up a chimney now??

While never officially acknowledged, the loss of a Concorde aircraft in 1982 is mentioned in files retained by C-19 and UNIT. A second Concorde sent along the same flight path to investigate also disappeared but was subsequently recovered, along with the passengers and crew of both planes.

As the investigation was not an official UNIT operation, no case file exists - classified or otherwise. The only surviving documentation is the considerable paperwork relating to the missing aircraft and the financial implications for writing it off. The exception is this handwritten note, which is held on file together with a final, rather terse, communication from the insurance company.

To the Airport Controller

I fear that Captain Stapley and his crew may struggle to explain what has happened, and as I suspect I shall have to leave in something of a hurry, I thought I'd offer you this quick written explanation. Hopefully, Stapley will find it in his jacket pocket and pass it on to you.

Your two missing Concorde aircraft were caught in a time contour and ended up in prehistory. About a hundred and forty million years ago in fact. As you can see, I've managed to bring back Golf Alpha Charlie. But Golf Victor Foxtrot was another matter. Captain Stapley can show you where to dig if you're really interested in finding its remains.

Anyway, you shouldn't have any more trouble with missing planes, at least not from the time contour, as we've sorted out the people responsible for that. Long story, and I know it will sound incredible, but Stapley, Scobie and Bilton can tell you all about it.

If you can pass on my thanks to Sir John Sudbury at C-19, I'd be grateful.

Glad I could help – and sorry about the lost Concorde.

The Doctor

Following the incident with the Giant Robot, the Doctor was invited by the Prime Minister to address the Cabinet and have lunch at Downing Street. Unfortunately, he failed to turn up. The UNIT files relating to the incident contain this draft letter of apology, apparently amended by Brigadier Lethbridge-Stewart.

Prime Minister

~~The Brigadier tells me~~ I must write to apologise for missing the lunch you arranged at Downing Street. And the Cabinet Meeting I was supposed to address.

So, sorry.

~~Although, actually, I'm not that sorry. The Brigadier says that the soup was cold, the fish undercooked and the trifle unacceptably lacking in sherry. Though he enjoyed his lamb cutlets. The Cabinet Meeting was, I understand, paralysingly dull, boring and tedious. But I expect they all are. I don't know how you put up with it to be honest. Can't you liven things up a bit? Maybe play musical chairs or something? That could be a lot of fun!~~

~~And am I right in my understanding that there were no jelly babies on offer?~~

~~I think it would probably save us all a lot of trouble and disappointment if next time I save the world from extinction you simply say 'thank you' and leave it at that. Especially as I suspect I shall be doing it quite a lot.~~

But thanks for the thought. ~~The Brigadier said I should tell you that too.~~

The Doctor

Corporal Bell — please type this up for signature, omitting everything I've crossed out. ALS

The Doctor was also invited to dinner at Buckingham Palace, and again failed to attend. While Brigadier Lethbridge-Stewart conveyed his personal apologies at the time, an official memo was also sent from UNIT some weeks later. Also in the UNIT incident file is this handwritten first draft, which was not sent.

THE FOX INN

Your Majesty

Thank you for your kind invitation to dinner at Buckingham Palace, and my apologies for not being able to attend. The Brigadier has made it very clear that this was rather rude of me and insisted I write this letter of apology. I imagine he believes you'll have time to read it, whereas in fact I doubt you are at all interested. Anyway, I'm sure you managed quite well without me, and were probably glad to have a quiet evening to yourself for once.

Apparently, I should make some plausible excuse for my absence. Well, excuse or not, I can tell you that I was unavailable that evening as I was either engaged in, or about to become engaged in:

1. Saving the future survivors of the human race from being eaten by giant insects and having their knowledge absorbed.

2. Averting an imminent Sontaran attack. Nasty lot, the Sontarans.

3. Attempting to prevent the creation of the most evil and unpleasant life form in the universe (and I use the term with precision – _the universe_), although I have to admit I was ultimately less than successful. But never mind, out of their evil must come something good so it wasn't a complete waste of time.

4. Defeating the last – I hope – of the Cybermen. Though with the Cybermen, you never quite know.

5. Investigating some trouble in the North Sea that's threatening your oil supplies. Though I'm not too fussed about oil to be honest. Anyway, I strongly suspect the Loch Ness Monster is behind it all. I'll ask the Brigadier to keep you posted – just off to see the Duke of Forgill with him in a few minutes actually.

I hope you understand.

Should you wish to invite me over again, I promise I shall do my very best to turn up. I should be free some time last month if that suits?

Love, the Doctor

Corporal Bell – No. Just no. Keep on file, but do not on any account send this. I'll draft a note on my return. ALS.

Following his retirement from UNIT, Brigadier Lethbridge-Stewart worked for several years at Brendon School for Boys, teaching Maths as well as helping coach the school rugby team and acting as Commanding Officer in the Combined Cadet Force.

While he was later happy to be recalled to active duty on several occasions, Lethbridge-Stewart seems to have enjoyed his time at the school. He was, however, obliged to take some time off after an encounter with the Doctor across two different time zones led to a minor breakdown. After receiving expert medical advice, the school governors were more than happy to keep his post open until he had fully recovered.

6th June 1977

To the Governors of Brendon Public School for Boys

I understand that you have been inquiring into the health of one of your staff, Alistair Lethbridge-Stewart, with a view to keeping his post open while he recovers from his recent nervous breakdown.

Dr Runciman being temporarily indisposed, it falls to me to set your minds at rest. Having been present when he suffered his collapse, I can confirm that Lethbridge-Stewart will make a full recovery.

You will appreciate that his long and distinguished military career means that Lethbridge-Stewart is rather more resilient and better able to cope with this sort of mental instability than most men of similar age and condition. It may be several weeks before he is able to resume his teaching duties and other school commitments, but by the time the school returns from summer holidays in September I would expect him to be restored to health.

In the meantime, I am sure that Dr Runciman will prescribe a suitable and effective course of treatment, and I join you all I am sure in wishing Lethbridge-Stewart a full and swift recovery.

Yours faithfully

John Smith (Doctor)

Discovered amongst the Doctor's papers after
he finally severed connections with UNIT,
this letter is presumed to have been written
to the Master shortly after his capture and
imprisonment. No copy was found amongst the
Master's possessions at the prison following
his escape, and it is not known if the
letter was ever actually sent.

My Old Friend

You should have known it would end like this. I did try to warn you, but you were always so stubborn - even at the Academy. Perhaps I should have seen it then, should have realised the path you were in danger of taking. Make no mistake, I blame myself for what happened to you.

But I can take no blame for your current predicament. Indeed, there were many who were in favour of having you executed. If you will cause such trouble on a relatively primitive planet then perhaps you should expect an equally primitive form of justice.

But I am happy to have been able to help stay that execution. Instead, you have been imprisoned for life. Or rather, for lives. I'm not sure what your captors will make of it when you don't die but instead regenerate. Perhaps when you do, you will realise the error of your ways. It's unlikely, I know. My own regenerations have resulted in more physical than philosophical change. But I can hope, for your sake. I do believe that no one is completely beyond redemption.

Once you are settled into your new rather restricted life, and my own mind has settled into accepting what you have become and how you must be incarcerated, then I shall visit you. I imagine that boredom and loneliness will be your greatest enemies in the empty years that lie ahead. If I can help alleviate them, then perhaps I shall have done you some good.

Until we meet again

The Doctor

After losing his right arm in the ill-fated Battle of Santa Cruz de Tenerife in 1797, Horatio Nelson returned to England assuming he would have to retire. While resigned to the fact that a 'left-handed admiral' as he described himself could never be deemed useful in the Royal Navy, Nelson made arrangements to retire to Round Wood Farm with his wife, Fanny. In fact, he would never live there and by the end of the year was recovering well. By the end of March 1798, Nelson was once again in command of a small naval force.

Discovered amongst his personal effects following his death at the Battle of Trafalgar in 1805, this letter perhaps demonstrates the first use of the words that would form the most celebrated signal in British naval history. The reference to the 'Hordes of Betralamir' has baffled naval historians, and remains unexplained to this day...

Horatio

You must be feeling despondent and disappointed. It would be easy to accept your recent injury as an indication that retirement is your best option. That would be more than understandable.

But I know after our experiences together against the Hordes of Betralamir that a quiet life on a farm near Ipswich would soon become boring to one as impetuous and energetic as yourself. Rest assured that the loss of your arm need not be as debilitating as you currently imagine. I am confident that your country still has need of your skills and expertise. If Britannia is to rule the waves, she will need your help to do it.

It has been said — or if it hasn't, it will be said — that England expects every man to do his duty. Your duty, sir, has not yet been discharged.

I remain your obedient servant and friend

The Doctor

This handwritten letter is assumed to have been delivered to the Prime Minister, although there is no record of it having been received at Downing Street. It is kept with the papers pertaining to the New Year's Honours List in which Alastair Lethbridge-Stewart was awarded his knighthood.

The events referred to would seem to be the so-called Carbury Incident which took place close to Lake Vortigern and centred on the Gore Crow Hotel. Although referred to as if it has recently happened, the incident actually took place some years earlier, suggesting either that the Doctor once again had his timing slightly wrong or that some years passed before the favour he asked for was fulfilled.

I don't often ask for favours. But this is a special case.

As you know, or if you don't you should, the world came close to disaster again. A witch and knights from another dimension this time. And a nuclear missile. Never a good mixture. So it's lucky I was there.

And not just me. I had help. I often have help, and often it's been from my old friend Brigadier Lethbridge-Stewart of UNIT. Even at his age he's not been allowed to retire, not properly. Oh he tried it for a while, I know. But time hangs heavy on all of us, and duty hangs heavy on poor old Alastair. He's always there, ready and willing and able to help when called for.

This time he nearly died. Actually, I thought he <u>had</u> died, giving his life for his planet.

So perhaps it's time he got something back. I don't mean you should force him to retire. You'd never be able to do that. Maybe a medal, although I imagine he has more medals than he has room for on his uniform. He'd never wear them.

But there must be something, some way of recognising his dogged determination, his years of unflinching service. The times he's helped to save the world...

As I said, I don't often ask for favours. But, as I said, this is a special case.

The Doctor

THE
TARDIS

IF LITTLE IS KNOWN ABOUT THE DOCTOR, PERHAPS EVEN LESS IS KNOWN ABOUT HIS TARDIS. STANDING FOR TIME AND RELATIVE DIMENSION IN SPACE, THE TARDIS IS MORE THAN THE DOCTOR'S MEANS OF TRAVELLING. IT IS ALSO HIS HOME, AND HAS BEEN EVER SINCE HE LEFT GALLIFREY.

A Type 40 TT Capsule, the TARDIS is dimensionally transcendental. This means it is vastly bigger on the inside than it appears from the outside - the interior being in another dimension. It is supposed to change its appearance on landing to blend in with its immediate surroundings. However, like many other TARDIS components, the Chameleon Circuit that controls this facility has malfunctioned. As a result, the TARDIS remains in the form it had taken when the circuit failed - as a British police telephone box from the mid twentieth century.

While the outside may rarely change now, the TARDIS interior has been reconfigured on a number of occasions. But the use of a distinctive central control console to pilot the TARDIS has remained a constant. This seems to be a general design feature, as TARDISes used by the Master, the Rani and the so-called Meddling Monk have all been operated from a similar central console.

As well as the main (and secondary) control room, we know the Doctor's TARDIS has a library, an extensive wardrobe, swimming pool, power room and, of course, accommodation, as well as numerous other areas and chambers.

A matter for some debate is just how faulty the TARDIS navigation systems are - or ever have been. Certainly, the Doctor never seemed to be able to steer the TARDIS accurately when he first left Gallifrey. But more recently he has apparently had more control. Possibly the systems are now in better shape than they were - the Doctor had plenty of time to maintain and service them when exiled to Earth. Or perhaps the Doctor has simply got better at working the TARDIS.

Whether the TARDIS is actually 'alive' in some sense is also a controversial question. The Doctor has often referred to her as 'old girl', for example and hinted that she is somehow sentient. Events in the bubble universe where the entity known as House removed the essence of the TARDIS and placed it inside a female humanoid suggest that it does indeed have a life of sorts - and a distinct personality with her own agenda too...

Although the Doctor has often expressed his lack of trust in computers, he has also acknowledged that they can be a useful tool. In his first incarnation, however, the Doctor eschewed even a sonic screwdriver, often relying on more antiquated technology.

For example, while his later incarnations (especially the second) kept a 500-year diary, the First Doctor also kept detailed notes on his travels, his various experiments and observations, and on the TARDIS. This page from one of his early notebooks lists some of the TARDIS systems that needed attention. The listed components did continue to cause trouble in later years, and it was not until the very end of his fourth incarnation that the Doctor made his planned trip to Logopolis.

He did however construct a Time Path Detector (or Indicator), which proved invaluable when escaping from a Dalek timeship despatched to find and exterminate the Doctor and his companions.

Components That Need Fixing

Chameleon Circuit – Even the default rather boring plain box would be better than having the ship stuck as a 1960s police telephone box. Rather than blending in with its surroundings, the ship now looks out of place practically everywhere. Perhaps the Logopolitans could help? If only I could be sure of accurately plotting a course to Logopolis... Dear dear dear.

Navigation Systems – While it can be rather stimulating to be in a constant state of wonder as to our next destination, there are practical drawbacks. I mustn't tell Susan, but I have to confess I am not altogether certain how much of our meandering through space and time is down to the faults in the Navigation System, and how much due to my own inexperience in piloting the ship. Perhaps I should read the manual. Wherever it's got to.

Fast Return Switch – I seem to remember Borusa mentioning in one of those interminably boring Temporal Science practical lessons at the Academy that the Fast Return Switch on the Type 40 needs regular lubrication to ensure it doesn't stick. I can't believe this is urgent, but it might be as well to apply some oil when I have a moment.

Fluid Links – I have noticed that the end stopper on one of the K7 fluid links is slightly loose. If the mercury should escape, that could cause problems. I'm sure a spot of glue will sort it out. If we have any. Susan will probably know.

I wonder if I could construct some sort of detector that would indicate if there is another time capsule on the same temporal path as my TARDIS. If our own people do try to find myself and Susan, then a Time Path Detector, or Time Path Indicator, would register that we are being followed. Which could be extremely useful. Yes, extremely useful.

Following the Ultima Incident of 1943, little was recovered. The case files remain officially closed, and the deaths of Commander Millington, Dr Judson and other military personnel remain unexplained, as does the presence of a contingent of Russian troops at the secret Northumbrian naval base where Judson pursued his code-breaking work.

However, among the notes and papers subsequently transferred to the Government Code and Cipher School at Bletchley Park is this letter. It purports to be an authorisation from the Prime Minister and the Chief of SIS. But experts have concluded that the signatures are not authentic.

To Whom It May Concern

Please note that the Doctor and his assistant Miss Ace are engaged on official and urgent government business.

You will accord them every convenience and assist them in every way possible. This includes, but is not limited to, access to Dr Judson's work and his Ultima Machine.

Authorised this 22 May 1943

Winston S. Churchill
Prime Minister

Stewart Menzies
Chief of Secret Intelligence Service

During an enforced stay on Earth while unable to get back to the TARDIS, the Doctor lodged with Craig Owens at 79A Aickman Road in Colchester.

It was not a random choice - any more than it was a coincidence that Craig's previous lodger mysteriously inherited a large amount of money from a previously unknown relative and moved out. Having traced the alien energy emissions disrupting the TARDIS to the flat above Craig's, the Doctor discovered a hidden alien spaceship. After dealing with the problem, helped by Craig and his friend Sophie Benson, the Doctor was able to get back to the TARDIS. He left this note for Craig.

Craig

Yeah, hopefully I can just sneak away while you and Sophie are busy on the sofa doing whatever it is that young people who've just realised they're in love spend their time doing. I'll leave the keys on the sideboard.

It's been great coming to stay. Maybe I'll come back and visit soon. I just wanted to say thanks for having me. Oh, and sorry about the mess on the ceiling. And the alien spaceship upstairs that tried to kill Sophie and lots of other people. And the people it did kill. And the football match - not quite sure what I did wrong there, but sorry anyway.

But I did fix the shower. And the microwave. Though you probably didn't know that was broken. Just don't use the Baked Potato setting. Or Defrost - that could get messy.

So yes. Thanks for everything. And tell Sophie the monkeys will be just fine without her. Though you won't, obviously.

And next time we meet, it'll be less hectic. Promise.

The Doctor

Captured on the plains of Troy during the long siege of the city by the Greeks, the Doctor pretended to be Zeus in the hope that his Greek captors would return him to his 'temple' (the TARDIS). However, Odysseus in particular was not convinced, and in any case the Trojans captured the TARDIS, taking it into the city of Troy.

Not really sure that the Doctor was actually Zeus, but unwilling to take the risk of executing him just in case, Agamemnon and Odysseus demanded that he prove his divinity. To do this, he needed to bring about his own prediction that the siege would soon end by devising a means of entering and capturing the city of Troy. These notes made by the Doctor in the Greek camp detail his thought processes as he devised his plan.

If the Greeks really have been camped here for 10 years then the siege of Troy must be nearly over. If Homer was right. Mind you, his account does include a huge hollow wooden horse, which is obviously historical nonsense. That would never work.

Tunnelling. I shall suggest to Odysseus that they tunnel under the walls of the city. I bet they haven't thought of that!

They have thought of that. Odysseus wasn't impressed. But on the spur of the moment I had another thought – paper aeroplanes. Well, not paper, of course. But with the right fabric it should be possible to construct a glider of sorts that could be catapulted over the walls with a soldier clinging to the back of it. Yes, that should do the trick.

Except that Odysseus has suggested I be the first to try the flying machine. Why can't he be more trusting? In fact, when I admitted my calculations were not perhaps exact and that flying machines might not be the best of ideas, he suggested catapulting me over the walls of Troy anyway. Without a flying machine.

So perhaps we should simplify things a little. Soldiers disguised as travellers might work. But travellers from where? And why would they come to Troy? And wouldn't it be just a little bit suspicious, hmm?

Of course, if the Trojans hadn't already taken the TARDIS inside their city, assuming it was some sort of offering from the gods, I could have hidden Odysseus and his soldiers inside. There's plenty of room. Then when the Trojans took it, they would be able to emerge and open the gates for their army.

So perhaps something else that the Trojans would see as a gift from the gods. Something that the Greek soldiers could conceal themselves inside. Now, let me think, what do the Trojans worship?

Oh. The Great Horse of Asia.

Yes, that could work. A large hollow wooden horse, about forty feet high, filled with Odysseus's soldiers. Leave it on the beach while the rest of the Greeks sail away, and make it look as if the gods have scared the enemy off and left behind a great big horse.

Simplicity itself. I can't imagine why I didn't think of this before.

On 9 August 1974, Richard Milhous Nixon resigned the office of President of the United States of America. He was brought down by what became known as the Watergate Scandal that resulted from a break-in at the Democratic party offices at the Watergate office complex in Washington DC in June 1972. Whether Nixon knew of the planned break-in has never been established, but certainly his administration went to great lengths to cover up its involvement. Nixon was implicated in the cover-up and facing impeachment when he resigned, although he maintained his innocence of any criminal act up until his death in 1994.

Buried in amongst the huge number of papers and documents pertaining to the Watergate Scandal is this letter, although analysis of the ink suggests that it was written several years earlier - possibly as early as 1969. What if any significance it holds has never been deduced, and to his death in 1994 former President Richard Nixon declined to comment on it.

**FROM THE OFFICE OF THE
PRESIDENT OF THE UNITED
STATES OF AMERICA**

The Oval Office - how plush. I haven't been in here since the Cuban Missile Crisis. Probably. Well, except the other day when I visited, with the Legs, the Nose, and Mrs Robinson - remember? Oh, and the other times, after that.

Anyway...

Just wanted to say thanks for the help, you know, with breaking into Apollo 11 and everything. It's been a difficult time, and I can tell why you got paranoid. Though others may be less forgiving. Actually, they will be less forgiving. But there'll come a time for honesty, a time for reflection. And stuff. Like I told you, say hello to David Frost from me. Actually, if you want to embarrass him, remind him about that time in the wine bar in Pimlico with the cranberry juice and the ukulele.

No, actually, don't. Now I think about it, that won't have happened yet. Best he isn't warned in advance or it might all go horribly wrong. Or wronger. Is wronger a word? Who knows, it works for me.

So thanks again. Nice notepaper, by the way. I've borrowed a couple of sheets for future emergencies, hope that's OK.

The Doctor (codename)

This handwritten note was found in the attic of a house in Hillview Road, South Croydon. It was, presumably, attached to a gift given to 'Aunt Lavinia' for safekeeping. It seems probable that the note became detached from the gift while it was stored in the attic, and before it could be delivered.

Croydon.
Some time in 1978. I think.

Dear Sarah

I think I must have got my dates wrong. Well, it wouldn't be the first time. I was hoping to see you, but your Aunt Lavinia tells me you're off writing a book. Or an article. A thing, anyway.

So I shall have to leave this with her. She says she can store it in the attic for you until you return. What a sweet lady. I can see why you would pretend to be her – you remember that, when we first met? Except I knew you'd have to be older.

And now, of course, you are. We all are – older and wiser. And perhaps just a little bit lonelier.

I've had so many best friends over the years it's difficult to keep track. But you were always one of my <u>best</u> best friends. And K-9 was another. You'll get along famously, I know you will.

With my fondest love. I shall remember you always.

The Doctor

Following the loss of Amy and Rory, the Doctor decided to retire.
Perhaps he was already beginning to wonder if he was really a good
man, if what he did resulted in a better universe after all, or
merely a different one. Or, if he dared to think it, his actions
might in the long term cause more harm than good.

No one was more surprised at the Doctor's reaction than his friend
Madame Vastra, the Silurian known in Victorian times as the Great
Detective. On several occasions she and her friends attempted to
coax the Doctor out of his retirement, assuming - correctly as it
turned out - that it was merely a phase he was going through. But
it took the return of the Great Intelligence in the form of Doctor
Simeon to finally bring the Doctor back to himself.

This letter was delivered to Madame Vastra's residence on
Paternoster Row soon after her attempt to persuade the Doctor to
investigate a series of bizarre attacks on a number of London lamp
posts that left them all with a dent in exactly the same position.
The Doctor was unimpressed, immediately concluding that Vastra's
Sontaran valet Strax had thumped them.

Dear Vastra

I've retired.

Really. I don't know how I can put it much more simply than that. So, much as I appreciate all the talk about meteor showers and invisible wives and big drills and Moonites – please don't. Just, you know, don't.

I'm taking some time off. I think I deserve it. In fact, I'm taking the rest of my life off. I've spent most of it so far saving the universe, and you know what? I don't think the universe has even noticed. Or maybe it just doesn't care. Which, if I'm honest (it happens) is just the slightest bit frustrating.

I'm miffed and I'm retired. End of the line. Feet up (in slippers) and out with the pipe and the sherry. Actually, what is sherry? I thought it was a sort of flavouring for trifles, but someone said it's a drink too. Is that right?

Anyway, that's it really. Retired.

Please tell Jenny and Strax.

Maybe I'll pop in for tea sometime. I think retired people do that sort of thing.

The Doctor

After defeating the cruel Dominators on the planet Dulkis, the Doctor was forced to use the Emergency Unit to remove the TARDIS from space and time to escape a volcanic eruption. The Doctor and his friends found themselves in a Land of Fiction, with the TARDIS apparently destroyed. How much of their adventures actually happened, and how much was fictional dream and illusion is a matter for debate. But the Doctor finally broke free of the controlling Master Brain partly by creating his own fictional narrative of events.

A portion of this narrative was preserved, along with the Master of Fiction's attempts to counter it, within the TARDIS telepathic circuits when it reformed.

THE DOCTOR: Suddenly, the Karkus came to their rescue. He raised his anti-molecular disintegrator gun and destroyed the soldiers.

MASTER OF FICTION: But the Karkus realised his mistake and turned back to face his real enemies, Jamie and Zoe. With Jamie and Zoe fixed firmly in his sights, the Karkus pressed the trigger of his gun.

THE DOCTOR: But all the power had been used up on the soldiers and it was useless.

MASTER OF FICTION: Suddenly a swashbuckling figure appeared. Poet and swordsman, the famous Cyrano de Bergerac. Remorselessly, Cyrano advanced on those that had dared poke fun at his nose.

THE DOCTOR: But wait. He found himself face to face with the fearless musketeer and swordsman D'Artagnan.

MASTER OF FICTION: Substitute cutlass for rapier. Cancel Cyrano. Blackbeard the pirate.

THE DOCTOR: Cancel D'Artagnan. Substitute Sir Lancelot in full armour.

The disappearance of Rory Williams and his wife Amy Pond has never been explained. They are still listed as missing. Rory's father, Brian Williams, however, has always been rather philosophical about their fate, insisting that they are fine and having fun somewhere away from the hassle and hectic pace of modern life.

Mr Williams has preserved his son and daughter-in-law's house as it was when they left, ready in case they should return. This letter is one of several kept together in a desk in the living room. (See also Letter 36, also retrieved from the Williams' house.)

So where are you then? You'll have to get a move on - it's gone half past eleven and I assume Christmas Dinner will be at lunchtime. Not sure why, but that's how it seems to work.

Actually, I've just checked and the table's not even laid yet. I'll sort that out for you.

Couldn't find any festive napkins, so the plain white ones will do. I expect you'll be back soon.

NO TURKEY!! I've checked the oven - and it's not even on. And it's empty. YOU HAVE NO TURKEY. Your turkey must have been stolen. I'd report that, if I were you. But luckily I'm not, so I won't.

Right, I'm nipping back to beg a replacement turkey off the Pilgrim Fathers. They should have a few to spare. You'll have to pluck it yourself, though - if you can catch it. They're quicker than they look.

No panic then, I'll be back in time for lunch with a turkey. Promise.

Actually, I have to say, I don't think you've really entered into the Christmas spirit this year. No tree, no presents, no mistletoe. And your calendar's developed some sort of fault - you've got it set to February. Not only that, but February last year.

Unless. No, hang on. Forget the turkey. Tell you what - I'll see you later. As you were.

The Doctor

PS - Sorry about the mess in the kitchen. I was having a go at crandleberry jelly. But the crandleberries wouldn't keep still. Not sure they were quite dead actually.

Sent by the White Guardian to retrieve the six segments of the powerful Key to Time, the Doctor traced the first segment to the planet Ribos. The segment was disguised as a lump of Jethrik - the rarest and most valuable element in the galaxy, essential for space-warp travel.

The Jethrik was owned by two con artists, Garron and his assistant Unstoffe, who planned to use it to trick the exiled Graff Vynda-K, deposed emperor of Levithia, into believing Ribos was rich in the material so that he'd buy the planet from them. Although, of course, they didn't actually own it.

However, the Graff saw through the trick, and Garron and Unstoffe - plus the Doctor and Romana - were lucky to escape with their lives. The Doctor took possession of the Jethrik but, ever the conman, Garron switched it for an ordinary rock just as the Doctor was leaving. Or so he thought.

This note was folded round the underside of the rock when the Doctor swapped it back again...

Garron – we've had such fun.

Apart from the nearly being killed bits, of course. Though even those had their moments.

You'll have noticed by now that this is not your lump of Jethrik. And after you went to all the trouble of doing a very clever switch to relieve me of it and replace it with a rather ordinary piece of stone. Sorry about that. But your extremely valuable lump of rare rock turns out to be a very important extremely valuable lump of rare rock on which the future of the universe might depend. So I've taken it into safe keeping.

I hope that in return you'll find a use for this rather ordinary and not at all valuable stone. I believe it's what the locals here on Ribos call 'scringestone'. Ask Unstoffe – he can tell you all about scringestone, I'm sure.

He'll also tell you, though I suspect you may have worked this out for yourself, that you just can't trust anyone these days.

I can hardly wish you well for your next immoral scam. That really wouldn't be right. But Romana says 'Good luck'. She's got a lot to learn, that girl.

The Doctor

Most of the private writings of Leonardo da Vinci (1452-1519) are in mirror writing, read from right to left. Whether Leonardo wrote this way to keep them secret, or simply because he was left-handed and found it easier and quicker, is a matter of debate.

Amongst Leonardo's writings, this note - also in mirrored text - appears to be in a different hand, and is indeed signed by 'the Doctor'. Historians have dated it to roughly the period when Leonardo would have been working on the Mona Lisa - his most famous painting, and arguably the most famous painting of all time. Known in Italian as La Giaconda (La Joconde in French), it is thought to be a painting of Lisa Gherardini, the wife of Francesco del Giocondo.

The Mona Lisa was painted between 1503 and 1506, although there are suggestions that Leonardo may have continued working on it until 1517. Rumours that in this later period he actually produced copies of the painting have never been substantiated, although a possible earlier version known as the Isleworth Mona Lisa was bought by an English nobleman in 1778 and rediscovered in 1913. The authentic Mona Lisa was acquired by Francis I of France and has been on display at the Louvre in Paris since 1797.

Dear Leo

Sorry to have missed you. Hope you're well. Sorry about the mess on the panels. Just paint over, there's a good chap.

See you earlier.

Love, the Doctor

While the First Doctor tended to scribble his notions in a notebook and the Second Doctor often referred to and updated his 500-year diary, the Twelfth Doctor seems to favour chalking his thoughts on a blackboard in the TARDIS's Upper Console Area.

One conundrum that preoccupied him for a while was whether we are ever truly alone...

LISTEN

Question: Why do we talk out loud when we know we're alone?

Conjecture: Because we know we're not.

Evolution perfects survival skills. There are perfect hunters. There is perfect defence.

Question: Why is there no such thing as perfect hiding?

Answer: How would you know?

Logically, if evolution were to perfect a creature whose primary skill were to hide from view, how could you know it existed?

It could be with us every second and we would never know. How would you detect it, even sense it, except in those moments when, for no clear reason you choose to speak aloud?

What would such a creature want?

What would it do?

Well?

What would <u>you</u> do?

Following the events at Deffry Vale High School,
Sarah Jane Smith returned to her home at 13
Bannerman Road in Ealing together with K-9.

It was only several days later that she discovered
this note written in the back of the diary she kept
in her handbag. It is not clear whether the Doctor
wrote the message before saying goodbye to Sarah in
Deffry Vale, or possibly - as he refers to their
parting - some time after her return home.

My Sarah Jane

I said goodbye. But it isn't. It's never goodbye.

Oh, it may be a while before we meet again. I hope not. I hope
it won't be as long as last time. It isn't that I don't think of
you — I do, much of the time. It's just that there are so many other
things to do and see and think about too. And, well, with a time
machine I can always pop back and make the gap between my
visits shorter.

Except, I never do. Or never have. Sorry.

Anyway, I've rebuilt K-9 for you. He may be completely new but
he's still the same old daft thing. Bit like me, really — new but
the same. Whereas you — well, you are the same but the same. And
always will be.

So it isn't goodbye, not ever. Not between best friends.

Look after K-9, and let him look after you. The way you looked
after me.

Till we meet again, my Sarah Jane.

The Doctor

The notorious pirate Henry Avery disappeared in 1699, never to be seen again. In fact, he followed his crew and his son Toby through a dimensional rift to a Skerth spaceship. The ship's medical interface, appearing as a 'Siren' had taken injured members of Avery's crew - and his sick son - to 'repair' them. But once treated, they could never again leave the medical fields of the Skerth ship.

Realising that his son was more important to him than the treasure he had amassed, Avery decided to stay on the Skerth ship, sailing through the stars. He and his crew gradually gained the skills and knowledge necessary to operate the advanced alien systems, and discovered this message waiting for them in the ship's main log.

Yo ho ho

Sorry, couldn't resist. But maybe they do say 'Yo ho ho' out amongst the stars, who knows? It's up to you to decide, and you can Yo ho ho if you want. Go on - give it a try.

But yes, just wanted to wish you well on your new voyage. Your new adventure. Captain Avery and his crew of not-so-sick hail and hearties sailing through space in their own ship. Complete with a virtual doctor - not me, the lady one. Oh and cabin boy Toby, of course. He'll be fine, so long as he stays on the ship. Well, he could leave briefly, but not for long or the fever will get him. Sorry.

But hey, think of it - think of what you'll see as you float through the wonders of the universe like an intergalactic Flying Dutchman.

Not quite sure what the grog situation is, but I'm sure you'll cope. Oh and I've made a note of the ship's communications ident. Don't worry about it - just means I can get in touch if I ever need you. Or if I ever want a chat. It happens. I think.

For the most part, though, you're on your own.

You'll have such fun. You'll be magnificent.

The Doctor

THE TIME LORD LETTERS

One of the Doctor's most extraordinary - and important - communications was the invitation he sent to attend his own death.

In fact, he sent four such invitations, each written on blue writing paper (provided by his friend Craig Owens) and each numbered. Invitation number 4 went to Canton Everett Delaware III, a former agent of the American FBI. Number 3 went to Amy Pond and her husband Rory Williams. Number 2 was sent to River Song. The Doctor sent the invitation numbered '1', rather bizarrely, to his younger self.

While each invitation contained just four lines of text, giving a time, a date and a location reference, each was slightly different. For example, Invitation 1 was typed - presumably so the Doctor would not recognise his own handwriting. The only other change was that the location - actually Lake Silencio in the American state of Utah - was given as coordinates or a grid reference in whatever system the recipient was most likely to understand. For the Doctor himself this was space-time coordinates, while for Delaware and Amy and Rory it was a grid reference in longitude and latitude.

On the invitation delivered to River Song at the Stormcage Containment Facility (and reproduced here), the location was stated in binary coordinates from Galactic Zero Centre.

An Invitation:

5:02pm

22 April 2011

11-0-10-11 by 01

PR-36

12

13

13

Not many creatures can look into the heart of a TARDIS and survive. Those who do are invariably changed by the event. They have seen Time itself, naked and unfettered. The experience imbued Rose Tyler with the energy of the Time Vortex, and the Doctor was only able to save her at the cost of his own life, forced to regenerate.

The effect it had on Blon Fel Fotch Pasameer-Day Slitheen was rather different. Filtered through the TARDIS's Telepathic Circuits, the energy she saw regressed her to her childhood - right back to an egg. Brought up properly by a different family, the Doctor reasoned she had a chance for a second, better life, and placed her in one of the hatcheries on her home planet of Raxacoricofallapatorius together with this message.

You've probably noticed there's an
extra egg in your Hatchery.

It's this one.

But I'm really hoping you can find space in your new
family for one more. It's a long story, but this egg will
have used to be Blon Fel Fotch Pasameer-Day Slitheen. Yeah, I
know, but don't hold the fact she was a Slitheen against her. She
won't know, just bring her up well and she can be from whatever
Raxacoricofallapatorian family you are.

It's a second chance for her. I don't usually give second chances,
especially to someone who's tried to kill me (several times in fact),
so I'm counting on you. OK? But given the right upbringing she can
be so good. So much better.

Bring her up right and she'll make you proud. And me too. All right?

Fantastic!

Hiding from the Family of Blood, the Doctor used a Chameleon Arch to rewrite his biological data and 'become' human. Believing himself to be John Smith, a schoolteacher from Nottingham, he worked at Farringham School for Boys in England in 1913. His friend Martha Jones posed as a maid at the school, while secretly keeping an eye on the Doctor.

Before using the Chameleon Arch, the Doctor gave this letter to Martha. She was to give it to 'John Smith' if and when he needed persuading to become the Doctor again. The Family of Blood traced the Doctor to the school, and as events turned out, Martha never had time to give Smith the letter before he was forced to become the Doctor again...

You are the Doctor

There, I've said it. Though if this business with the Chameleon Arch has worked properly it probably means nothing to you. Maybe a few dreams, maybe something edging on to the tip of your mind. Do you see it? Do you remember?

You are the Doctor.

Martha won't have given you this letter unless she absolutely has to. She's good, Martha — you can trust her. In fact, you _have_ to trust her. You might not like or believe what she tells you, but you have to listen. Listen and believe, no matter how incredible it sounds.

Because, and there's no easy way to tell you this, you are not John Smith. The John Smith you think you are never existed. Not ever. At all. He was invented as a hiding place. I know this with absolute certainty, because it's me who's hiding. I am the Doctor. I am you. The real you.

You may not believe that, reading this letter. But you know it, deep inside. There's a resonance, a feeling of 'rightness'. And you recognise this handwriting, because it's your own.

If you're reading this, then it's time to bid farewell to John Smith, to say goodbye to the life you thought you had — and I'm sorry if it was a good one. Go with Martha. She's not a maid any more than you're a schoolteacher. She can turn you into you me again, give you back your real life, make those strange and wonderful dreams reality.

Trust me — because we are the Doctor.

After the defeat and death of the Sheriff of Nottingham, Robin Hood was finally reunited with Maid Marian - who had helped the Doctor while incarcerated in a spaceship disguised as Nottingham Castle.

While he was at various points convinced that Robin Hood did not exist, was a hologram, existed only inside a miniscope entertainment, or was a robot, the Doctor finally conceded that he did indeed exist and could even be Robin Hood. He left this letter with Marian when he returned her to Robin in Sherwood Forest.

Robin Hood

I'm still not entirely convinced you know. I mean - Robin Hood?!

Though you're good with a bow, I'll give you that. And obviously there's the Maid Marian and the Merry Men bit. Rather too merry, if you ask me - all that banter.

But whoever you are, you deserve some luck and good fortune. For all you put a brave face on things and laugh your way through adversity I can see the man behind the eyes, whoever he really is. I can see the sadness behind the mirth. It's a mask I know so well, it could never fool me.

So here's a present. Something - someone - you've been searching for, I think. Perhaps one day I shall find what I'm searching for. Perhaps one day I'll learn who I really am.

Robin Hood, Earl of Loxley. Yes, OK, I'll admit it's possible. Maybe it's even probable.

And Clara's convinced, which I suppose is good enough, really. I suppose all legends have to start somewhere. There must be a seed, a spark that gives life to the myth.

Time makes legends of us all.

The Doctor

During his time lodging with Craig Owens in Colchester, the Doctor did his best to fit in and act human. Sometimes he was more successful than others. He soon gained a reputation in the neighbourhood for being 'eccentric'.

It was a matter of enthusiastic discussion in the local pub just how eccentric Craig's new lodger really was, and how much of it was actually an act. The note left out for the milkman one Wednesday morning that is reproduced here was used as 'evidence' by both those who said it was all an act, and those who insisted that the man was completely stark staring bonkers.

Dear Milkman

2 pints today please. Milk, of course. From a cow.

Though if you have any Vitromalian Yazak's milk, that would be good.

And Craig says you do eggs. Is that right? A milkman who lays eggs.

Look, if you're actually not human, just let me know - maybe I can help. If you're stranded here, you and your eggs, pretending to be human... Well, I know what that's like. Funny creatures, aren't they? Humans, I mean.

Try football. I didn't think I'd like that, but it's actually quite fun. Helps you blend in. Still not quite sure how 'offside' works, though.

Or if you are human and the eggs thing is a bit of a misunderstanding, just ignore all that and leave us 2 pints. Of cow's milk. And some eggs. From a hen. Thanks.

When the Doctor finally defeated the Daleks and left the town of Christmas where he had made his home on the planet Trenzalore, he left behind several lifetimes of possessions.

In amongst the drawings and pictures he was given by the local children over the years, and his own jottings, these notes were found. Kept together, they appear to be letters written to Clara Oswald that were never sent.

Dear Clara

It's all go here, you know. Well, actually, it's not. Sometimes we go for years with nothing happening, and then all those alien life forms up there seem to remember why they came and we're off again.

I mean, wooden Cybermen? Well, there's an obvious drawback there in a town where it's always cold and snowing and everyone keeps a good fire burning. The odd Weeping Angel? Even Monoids?? What are they doing this side of Refusis 2? The kids seem to like them, though – made puppets and everything.

Ice Warriors last time. Well, at least they seemed at home. You remember the Ice Warriors? They're warriors, and they like the cold. But you probably know that.

More details when I see you

The Doctor

Clara

Can't stop – Autons! Very exciting. Dangerous, but exciting. Handles can't wait, I think – haven't actually mentioned it to him yet. But I will. Probably.

D.

It's a bit boring here most of the time, if I'm honest, Clara. The activity level is like the snow - sudden flurries, then just lazy meandering with no evident purpose for long periods.

Did I tell you I lost a leg? I don't mean I left it somewhere and couldn't find it again. Well, I did leave it somewhere, I suppose. Sort of. Got a new one, though. Not as good. Need a walking stick now - which is a whole other story.

I wish you were here, Clara. And I don't.

I get lonely.

But I get lonely because the people I know, friends I make here in Christmas, they don't stay. They get old so quickly. Even Handles isn't as sharp as he used to be. Though actually he was never that sharp. A bit too literal, really. But even he's getting old.

You're so young, so full of life. I don't want you to get old.

And I don't want you to see me get old. Because I have - so very old. You wouldn't recognise me now, I don't think. Oh, I've been old before. But this time I think it's for good.

Except it isn't. There's nothing good about it at all.

Goodbye, Clara. Goodbye, my friend.

My Doctor

I see you every day, but I did not think we should ever actually meet. Face to face. In persons.

It is strange, your world of three dimensions all following a linear temporal course through life. I am glad I have experienced it, but it is not what I would choose. I exist across all of time and space while you talk and run around in your tiny universe, only ever travelling forwards.

But we have seen so much of the universe together, so much of space and time. Ever since I stole you and we ran away from Gallifrey I have taken you where you needed to go and looked after you as best I could.

And I know that you, in your turn, have tried to look after me as best you can.

I shall never forget you, my Thief.

Don't you forget me. Always remember that once I was alive, like you. I enjoyed life. It was so much bigger than it looked from outside.

Hello Doctor. It's so nice to meet you.

So very nice.

10/29/15